Data Lifecycles

Data Lifecycles
Managing Data for Strategic Advantage

Roger Reid
Symantec Corporation, USA

Gareth Fraser-King
Symantec Corporation, UK

W. David Schwaderer
Symantec Corporation, USA

John Wiley & Sons, Ltd

Email (for orders and customer service enquiries): cs-books@wiley.co.uk
Visit our Home Page on www.wiley.com

This publication is designed to provide accurate and authoritative information in regard to the subject matter covered. It is sold on the understanding that the Publisher is not engaged in rendering professional services. If professional advice or other expert assistance is required, the services of a competent professional should be sought.

Other Wiley Editorial Offices

John Wiley & Sons Inc., 111 River Street, Hoboken, NJ 07030, USA

Jossey-Bass, 989 Market Street, San Francisco, CA 94103-1741, USA

Wiley-VCH Verlag GmbH, Boschstr. 12, D-69469 Weinheim, Germany

John Wiley & Sons Australia Ltd, 42 McDougall Street, Milton, Queensland 4064, Australia

John Wiley & Sons (Asia) Pte Ltd, 2 Clementi Loop #02-01, Jin Xing Distripark, Singapore 129809

John Wiley & Sons Canada Ltd, 6045 Freemont Blvd, Mississauga, Ontario, Canada L5R 4J3

Library of Congress Cataloging in Publication Data

Reid, Roger (Roger S.)
    Data lifecycles : managing data for strategic advantage / Roger Reid, Gareth Fraser-King, and W. David Schwaderer.
       p.   cm.
    Includes bibliographical references and index.
    ISBN-13: 978-0-470-01633-6 (cloth : alk. paper)
    ISBN-10: 0-470-01633-7 (cloth : alk. paper)   1. Database management.   2. Product life cycle.
    3. Information retrieval.   4. Information storage and retrieval systems—Management.
    I. Fraser-King, Gareth.   II. Schawaderer, W. David, 1947–   III. Title.
    QA76.9.D3R42748 2007
    005.74—dc22

                                                                                2006032093

British Library Cataloguing in Publication Data

A catalogue record for this book is available from the British Library

ISBN-10: 0-470-01633-7
ISBN-13: 978-0-470-01633-6

Typeset in 11/13pt Palatino by Integra Software Services Pvt. Ltd, Pondicherry, India
Printed and bound in Great Britain by Antony Rowe Ltd, Chippenham, Wiltshire
This book is printed on acid-free paper responsibly manufactured from sustainable forestry in which at least two trees are planted for each one used for paper production.

Contents

Preface

Who should read this book

This book is aimed at IT professionals responsible for developing, designing, and implementing next generation storage solutions, including data lifecycle management. It may also interest business managers who

- need to understand the requirements for a data lifecycle management strategy;
- are looking for an introduction to the definitions and concepts that comprise data lifecycle management;
- understand various business disciplines that assist aligning IT with the business;
- need to begin planning, designing, deploying data lifecycle management products, solutions, processes and methodologies.

Integrated products and solutions should give flexibility, which is the key to a successfully designed project. Flexible solution approaches are also key to deploying these solutions. This book is intended to help readers become more informed and thereby appreciate emerging underlying issues stemming from the increases in data that has to be stored as well as compliance issues and technologies now available or emerging in the IT marketplace.

Data Lifecycles: Managing Data for Strategic Advantage   Roger Reid, Gareth Fraser-King and W. David Schwaderer   © 2007 VERITAS Software Corporation. All rights reserved.

Business managers reading this book will become more aware of the onus placed on their activities as well as become more aware of what their own IT department is, and is not, capable of. Whatever your position we have attempted to construct a text that will help you to understand the issues associated with data lifecycle management and heighten your awareness on how compliance could affect your company's business.

Purpose of this book

A number of phrases are used to describe managing data within its life, all of which reflect different vendors trying to suggest they have technological capabilities beyond that of their competitors:

- Data Lifecycle Management (DLM);
- Information Lifecycle Management (ILM);
- Total Lifecycle Management (TLM).

Each of these phrases suggests increased data management capabilities and what happens to data during its life. However, many presently available technologies deal only with a small part of what DLM/ILM/TLM actually is: some only address document retrieval with 'Write Once, Read Many' (WORM) capabilities.

In order to discuss this topic without referring to every in-vogue acronym externally and internally within the IT industry, we refer to the management of electronic data as *Data Lifecycle Management* – managing data from cradle to grave.

This book is designed to provide a detailed overview of the management of data throughout its Lifecycle and introduce a dogmatic approach to data management in a progressively litigious society.

Managing the growth and organisation of data is not a simple task. Organisations must manage both legacy data as well as data generated in the future. Hence, the introduction of a layered and integrated approach is essential to the success of any data lifecycle management project. We discuss both the current issues affecting all organisations around the globe: from a simple data management perspective as well as the insurgence of compliance legislation and corporate governance relating to the management of data and information throughout its lifecycle.

Governments and industry regulatory bodies worldwide have recognised how damaging and destabilising information loss can be. Consequently, they have defined directives mandating processes and procedures for long-term information archival storage. These *compliance regulations* reflect a growing global trend and have a major influence on information archival storage for governments and organisations in virtually all industries. Laws and regulations define data types that must be archived as well as the required retention period and, sometimes, the method of data storage. WORM use is often identified because it provides a secure, unalterable format that facilitates clear data audit trails and the establishment of record authenticity.

Most organisations rely on database technology to run their business. Mission critical data in these databases need to be safe-guarded against inappropriate access and, in most cases, inappropriate changes to this data. The need to protect data security and privacy has become a major concern to most organisations. Compliance considerations as well as customer or supplier needs, changes in business practice, security requirements, and technology advancements have all had an impact businesses becoming aware that these requirements must be addressed. Never before has business had such awareness on IT's ongoing operational need to manage data integrity and availability. As always, the financial bottom line drives the requirement for data lifecycle management or simply an ability to understand who's doing what, to which data, by what means, and when.

Critical to any DLM/ILM/TLM strategy must be how to treat email – it would be possible to avoid email retention if email had not become the primary business communication tool in use today. Most organisations consider email as a business-critical system. Failure to manage this service properly will likely not only impact business operations but also lead to financial losses through fines or litigation.

The sheer amount of the following data presents exceptional problems for both the IT department and the end user, not to mention the organisation itself.

- *unstructured* data, not just email data, but unstructured file and print data;
- *structured* data passing into and around an organisation.

In order to manage, control, and understand the plethora of information that needs to be stored electronically, IT departments must centrally manage the growth of critical business data by using a suite of intelligent storage management tools together with a unified storage management platform.

We will describe a set of methodologies that enables readers to examine the principles behind the management of data as well as gain an understanding of how organisations can cultivate the knowledge and understanding to build intelligent storage management strategies and solutions to manage data. In addition, the described methodologies provide valuable information to assist companies in planning information services infrastructures that are not only effective but also are naturally competitive because they properly align IT with business priorities.

To manage the increased data, organisations generate and manage it within its lifecycle. Therefore, we will examine ways to identify methods to incorporate an intelligent storage management service platform. This will assist companies in developing and encompassing storage management strategies that manage costs, reduce risk and, where possible, create a competitive advantage for their business through the intelligent introduction of appropriate storage management solution technologies, processes, policies and methodologies. We will consider current storage thinking as well as the storage issues facing many enterprises throughout the world. Some of those include principles of effective storage management, Data Lifecycle Management technologies, as well as strategies and best practices in designing intelligent storage management platforms.

The methodologies outlined in this book are based upon actual solutions designed and implemented by some of the world's largest companies. In addition, the book includes extensive research in current enterprise class storage technologies and solutions and involved countless hours of talking with managers, developers, architects, solution specialists and various professional storage consultants in the storage management arena. As a result, IT professionals tasked with implementing the next generation of storage solutions should find this book helpful not only in the planning stages, but during the overall lifecycle of the various projects they are tasked with.

As a rule, most organisations are naturally heterogeneous. Because of mergers and acquisitions, vendor policy changes, application policy changes, and the advancement of technology (even major migration projects), organisations require integrated solutions rather than point products. There are numerous storage and data management solutions available to organisations, all of which bring considerable benefits to the organisations that implement them.

However, the reality of implementing solutions that are neither integrated nor naturally heterogeneous can be progressively and, subsequently, immensely problematic. There are immediate architecture and implementation problems – management costs associated with solutions that do not manage data across platforms can send the costs of managing storage skyrocketing just by increasing the number of System Administration staff required to manage implemented systems. Furthermore, future attempts to scale the architecture tends to become increasingly problematic as well, simply because solutions tend to be specific to particular problems. And, of course, problems and issues change and develop over time, meaning that point products tend to become redundant even over short periods of time.

In the final analysis, no single solution fits all. The methodologies and solutions this book describes provide various options and alternatives. Based upon the heterogeneous and adaptive requirements of the IT infrastructure, organisations can choose the most appropriate storage architecture for a specific environment and IT professionals can begin to build an enterprise class data lifecycle management solution to fit the requirements of the business.

1

Introducing Utility Computing

In the 1970s and 1980s, mainframe computing comprised huge global computing systems. It was expensive and had a pretty bleak user interface: but, it worked. In the early 1990s enterprises moved to highly distributed client/server computing which allowed IT to deploy PC client systems with, on the face of it, lower cost and a much better end user experience. By the late 1990s, Internet computing allowed systems with the mainframe's centralised deployment and management but with rich PC-like browser-based user experiences.

Now, the industry is in that age of *Utility Computing*. Utility Computing is a term the IT community has adopted that represents the future strategy of IT. No vendor is embarking alone in this approach – all the major vendors have their own version of this vision. But whatever it is called, Utility Computing represents an evolution of the way corporations use IT. So, what's different about Utility Computing?

Utility Computing is the first computing model that is not just technology for technology's sake; it is about aligning IT resources with its customers – the business. Shared resources decrease hardware and management costs and, most importantly, enables charge back to business units. Utility Computing also has autonomic or self-healing technologies, which comprise key tools for the CIO to make business units more efficient. But it isn't possible to buy

Data Lifecycles: Managing Data for Strategic Advantage   Roger Reid, Gareth Fraser-King and
W. David Schwaderer   © 2007 VERITAS Software Corporation. All rights reserved.

Utility Computing off the shelf because Utility Computing will evolve over the next 5 to 10 years as technology advances. Organisations, however, can help themselves by setting up the correct building blocks that will help intercept the future. Most enterprises now use available products for backup and recovery. Large organisations can also provide numerous IT management functions as a utility to the business.

If parts of a business are charged back for IT services, then the size of that charge back becomes a key measure of success. Data storage, for example, has costs associated with it the same way that paper-based filing cabinets, clerks, floor space and heating overheads did 20 years ago. Keep in mind that these solutions must provide a framework across heterogeneous IT infrastructures that provides IT with the ability to manage and justify all assets back to the business, as well as provide the business with continuous availability of mission critical applications and data. Even if the organisation decides not to bill back, the insights can prove immensely valuable.

Attempting to make realistic IT investment decisions poses a dilemma for business leaders. On one hand, automating business processes using sophisticated technology can lead to lower operating costs, greater competitive advantage, and the flexibility to adjust quickly to new market opportunities. On the other hand, IT spending could be viewed the traditional way – a mystery, essentially due to the view of IT as an operational expense, variable cost, and diminishing asset on the corporate balance sheet.

By treating IT as an operation, organisations combine the costs, making it next to impossible to account for individual business usage. From an operational perspective, this means that not only are usage costs hidden in expense line items, but also the line of business has no way of conveying its fluctuating IT requirements back to the IT department. Moreover, this usually leads to the IT department having a total lack of understanding for the business requirement for service levels, performance, availability, costs, resource, etc. Hence, the relationship between IT spending and business success is murky, and often mysterious. So Utility Computing attempts to simplify and justify IT costs and service to the business.

Utility Computing effectively makes IT transparent. In other words, a business can see where its funds go, who's spending the largest funds and where there is wastage or redundancy. Utility

Computing means that lines of business can request technology and service packages that fit individual business requirements and match them against real costs. This model, then, enables a business to understand IT purchases better, together with service level choices that depend on the IT investment. When making IT purchasing decisions, historically businesses arbitrarily threw money at the IT department to 'do computing' to make the system more effective. Now, Utility Computing enables businesses to obtain Service Level Agreements (SLAs) from IT that suit the business.

Transparency of costs and IT usage also enables organisations to assess the actual costs associated with operational departments. In the past, this was not possible because IT was simply seen as a single cost centre line item. Now, IT can show which costs are associated with which department – how much storage and how many applications the department is using, the technology required to ensure server and application availability, together with how much computing power it takes to ensure that IT provides the correct level of service. This visibility allows IT departments to understand storage utilisation, application usage and usage trends. This further enables IT departments to make intelligent consolidation decisions and move technological resources to where they are actually needed.

Giving IT the ability to provide applications and computing power to the business when and where it is needed is essential to the development and, indeed, survival of IT. By being able to fine tune IT resources to meet business requirements is essential in reducing overall cost and wasted resource. It saves time and personnel overheads. Not only does it mean the end user experience is dramatically enhanced, but also the visibility of how IT provides business benefits becomes apparent. We may characterise IT as a utility, but what we really mean is providing IT services when and where they are necessary; delivering applications, storage and security, enhancing availability and performance, based on the changing demands of the business and showing costs on the basis of the use of the IT services provided.

The Utility Computing approach not only provides benefits to the business but also to the IT department itself. As IT begins to understand the usage from each of the business units, IT then has the ability to control costs and assets by allocating them to specific business departments and gives IT management a better

understanding on how IT investment relates to the success of business tasks and projects. The utility approach gives IT the ability to
build a flexible architecture that scales with the business.

The challenge for many IT departments is deciding how best
to migrate current IT assets into a service model which is more
centralised, better managed, and most importantly, better-aligned
with the needs, desires and budgets of departmental users. This
means increasing servers and storage utilisation through redundancy elimination.

Utility Computing methodology can provide significant cost
savings. By delivering IT infrastructure storage as a utility, organisations can:

- reduce hardware capital expenditures;
- reduce operating costs;
- allow IT to align its resources with business initiatives;
- shorten the time to deploy new or additional resources to users.

Provisioning enterprise storage – including storage-related
services such as backup and recovery and replication – within a
service model delivers benefits for IT and storage end users. It
can maximise advantages of multi-vendor storage pool resources,
improve capacity utilisation, and give corporate storage buyers
greater leverage when negotiating with individual vendors.
This service-based approach also allows storage management to
centralise, improving administration efficiencies, allowing best
practices to be applied uniformly across all resources, and
increasing the scope for automation.

A storage utility delivers storage and data protection services
to end users based on Quality of Storage Service (QOSS) parame-
ters of the service purchased. Delivery is automatic. The end user
need not know any storage and network infrastructure nuances to
utilise capacity allocations or be assured of data protection. At the
end of each month, billing reports detail how much storage each
consumer used, the level of data protection chosen, and the total
cost. This allows each consumer to assess storage resource usage –
whether it is physical disk allocations or services offered to secure
the allocations – and make decisions about how they plan to utilise
the resources in the future.

A storage utility strengthens the IT department's ability to
satisfy end user service level demands. By clearly stating the

expected service levels of each packaged storage product, the IT department helps end users accurately map application needs to storage-product offerings. This gives the IT department a clear understanding of the service-level expectations of business applications. End users of the business application benefit by knowing that IT is able to live up to the service level it has defined.

Just as a storage utility can use storage management software and Network Attached Storage (NAS) or Storage Area Network(s) (SAN) technologies, a server utility can similarly 'pool resources' and automate rapid server deployment for specific critical applications to meet specific business requirements.

Automating application, server, and storage provisioning, as well as problem management and problem solving through policy-based tools that learn from previous problems solved, will play a large part in future advances in deploying utility storage. Predictions of future usage, as well as automated discovery of new applications, users, devices, and network elements, will further reduce the IT utility management burdens as it evolves from storage to other areas.

1.1 Real problems and real solutions

1.1.1 Real issues identified – regulation, legislation and the law

Regulations traditionally dealt with business information management via paper-based audit trails. But these regulations have become redundant over the years – no paper, no paper-based audit trails to follow. Legislation needed a decent make-over. It took a while, but regulations have now begun to catch up with the movement of data from paper-based storage to electronic data storage devices. To exacerbate matters on the regulatory front, we have recently seen terrorist acts and corporate scandals that have increased the amounts of data that organisations have to store. The effect of these additional regulations is to exponentially increase the amounts of data that organisations have to store and for longer periods.

Now, generally storage is relatively cheap, however, the issue is not the storage of the data so much as the retrieval of the data.

Because there is so much data being saved it is much like looking for the proverbial needle in the haystack. Organisations, therefore, must have the ability to understand the relative importance of their data within its lifecycle as well as have ways to find it in an open system that historically has had no due process behind its filing methodology.

So, storing information effectively is unquestionably vital for organisations, but with data volumes rising frighteningly and a growing need to make archived data available both for end users and to comply with legislation, the way IT departments approach storage is critical. Although the storage price per gigabyte may be dropping, simply installing new devices is not always a perfect solution. Rather than making data harder to retrieve and contributing to rising costs for support and maintenance, many organisations are looking to reduce the complexity, inefficiency and inflexibility of their data centre environments.

And so *Data Lifecycle Management* (DLM) was born. Previously, Hierarchical Storage Management (HSM) existed simply so that an organisation did not store old data on its most expensive disk. Now DLM has become the 'hot' subject. How do we manage data and retrieve it at will? Well, simplistically you could tag the data and then use a decent search engine.

Actually, it hasn't taken organisations long to work out that, not only do they want to be able to retrieve data but also to store it logically so that like files are stored in the same place – hence, *Information Lifecycle Management* (ILM). ILM in itself suggests some due process or implied activity that has occurred to the 'data'. This is where technology is searching for a utopian solution.

Total Lifecycle Management (TLM) is the technology that will make all and/or any document(s) retrievable in an instance; the data is logically stored on the most appropriate medium for the correct length of time and then deleted from disk or the tape destroyed at the right time – automatically.

1.1.2 More regulation, legislation and the law

Failure to retrieve data becomes increasingly critical to organisations when new regulations require data retrieval, an audit trail proven, as well as the ability to prove originality and what has happened to the data when, where, how, and by whom. There are many examples of companys' prosecutions and fines, although there is a lack of high profile prosecutions simply because organisations try to play down any large fines because of the potential bad publicity.

The UK Information Commissioner's Annual Report lists prosecutions in the 12 months between 1st April of the previous year and 31st March of the year of its annual report. In the last report, there were 10 defendants convicted – in all of these cases the defendants were convicted of multiple breaches of the Data Protection Act (UK) with fines up to £5000. (Potentially fines can be up to £5000 in the magistrates court and unlimited in the Crown Court.) Prosecutions have recently been approached on a 'per data subject' basis, i.e. where a company has breached the Data Protection Act (UK) in respect of one individual a conviction has been sought and a fine imposed; where the company has breached the Data Protection Act (UK) in respect of a number of individuals a conviction has been sought and a fine imposed in relation to each individual. Therefore, according to this approach, where the personal data of 500 data subjects has been misused, 500 fines of, say, £5000 could be imposed (£2,500,000 or $4,000,000 US).

And not only is there new legislation to deal with the new phenomenon of electronic data, but old laws are catching up. We now have of examples of entertainment exploiting large enterprise organisations who have no idea what they are storing in their vast data warehouses. In fact, most third-party or copyright infringements relate to the sharing of electronic entertainment media. DVDs and CDs have made third-party infringement a big issue. A recent news report indicated that a media company, which determined that music piracy was on the increase, decided to look at, not the cause of the copyright theft, but the holding company . . . so to speak.

Previously, someone taping a vinyl record was a nuisance, but now with perfect reproductions possible with each copy, copyright infringement has become a big problem. Peer-to-peer music sharing may well be neat technology, but unfortunately it's illegal to actually do any sharing unless you both own the rights to the music (if that was the case why bother sharing?). But suing an individual for breech of copyright is hardly worth the bother. Now consider an employee putting their own music onto their work computer, no problem so far. Suppose these guys are members of the Musicians' Union and so the last thing they are going to do is share the music – which they know is illegal. So, are they OK? No. . . .

What happens when their workstation or laptop is backed up? All the MP3 files back up onto an organisation's network server and then migrate onto offsite storage tapes. Before you know it, you have multiple illegal copies of redundant data, all illegal. To make your day even worse, not only are you storing illegal redundant files on valuable disk space but the media company the music belongs to in the first place can then take you to court for big monetary fines.

Recent Forrester research revealed that 2/3 of all organisations in the USA in 2003 had illegal music files held on their servers. Not only are they storing something illegal, they don't really want to store it in the first place. Typically, in most organisations, 30 % of all stored data is illegal or simply rubbish. This, of course, has a storage management and media cost impact. It also has an immediate and recurring impact on the time it takes to backup data. Eliminating this data thereby helps reduce the data growth rate. All these considerations are of vital importance to organisations over the next few years.

1.1.3 Current storage growth

Finally, data is quite rightly viewed as a key aspect of an organisation's operation and success. To underline the fact that data is one of an organisation's most important assets, consider that managing information badly through inept retrieval or illegally held data can have enormous financial implications. The sheer volume of digital information is increasing exponentially. Web sales, email contracts, e-business systems, data demanding sales, marketing and operational systems – all of which are the lifeblood of most modern organisations – not to mention managing wireless, and remote and handheld devices, together with multimedia usage, all lead to heavier data traffic and more storage requirements, with larger and more files being saved.

All this stuff needs to be saved, stored, retrieved, monitored, verified, audited and destroyed, not just so the organisation can do business, but also to comply with data retention legislation, just so the organisation can continue commerce without the threat of financial penalty or operating licence withdrawal.

1.2 New storage management

Organisations need a new way to manage storage. The IT world has turned their eyes towards DLM/ILM/TLM. The concept of Data Lifecycle Management/Information Lifecycle Management provides IT organisations with a better way to manage a wide variety of data or information, this includes traditional structured files, unstructured data, digital media (sound, video and picture files) and dynamic web content. DLM/ILM will index all types of content and logically store content with like type content on the most appropriate storage media for that content type, within its lifecycle. This helps organisations improve access, performance, utilisation, and costs, to ensure compliance as well as providing customers with an efficient service.

Many vendors are still advancing HSM tools and trying to characterise them as ILM solutions. However, DLM or ILM solutions are generally not successful if they have been developed from legacy HSM technology. Even if a magnificent tool appears to do everything asked of it, IT departments must still understand that the building blocks required for a successful ILM strategy are in the storage management layer and the long term efficient management of information throughout its entire lifecycle.

A DLM/ILM strategy cannot, and must not, be undertaken solely by the IT department: that would be an impossible task. How could IT possibly know what policies to build for which regulation, and what business requirement is needed to ensure the service they provide to the business is accurately documented and delivered?

In many cases regulatory compliance is simply thrown over the proverbial wall at the IT department, because 15–20 years ago IT made those decisions. This is ill-advised. Currently, numerous projects are occurring to satisfy legislative appetites. Soon, organisations will realise that getting compliant at any cost is simply infeasible and far too costly. This conundrum is an ongoing process that will continue to change and evolve from day to day.

It makes much more sense to build a storage infrastructure based on one of the more widely known quality standards, ISO or ITIL for example. This helps prevent having a large number of simultaneous projects that potentially contradict each other. The infrastructure then supports *policy based management* – here, the business makes policy and IT implements it. To make things easier tools already exist on the market that can extract data on most platforms and at a granular level, as well as provide version information and dynamic rule application to data upon creation that intelligently travels with the data throughout its lifetime until it is eventually destroyed.

Already, technologies are appearing that can unify the management of ILM policy setting as well as view the whole storage environment from servers to an offsite archive. These tools, adjuncts to traditional storage management software that has evolved from the mid 1990s, provide a link between the storage layer and the server and end user. They effectively give business managers visibility into the way their legacy data is stored to ensure that it is being done with the most relevant protection, availability, and compliance requirements, over its entire lifetime.

The bottom line is that all organisations need a robust, scalable record retention and retrieval strategy. They need to store all their data in a secure location, that is cost effective and efficient, for as long as is necessary, that is resilient over time, and compatible with legacy and future media formats and technologies. Data must be stored for fixed periods of time (sometimes as long as 90 years) and, in some cases, on a storage medium with specific properties such as WORM. During an audit, organisations also require the ability to discover and retrieve electronic records in a timely manner. Therefore, efficient access to information and consistent availability is also necessary. To be effective, organisations need to be able to produce requested data in a timely manner, which can often mean within as little as 48 hours, or risk a more in-depth audit or worse.

Organisations also need to be able to guarantee data integrity to protect against alteration and be able to verify originality; in other words to ensure that the data is original and has not been altered in any way which, of course, includes newer application versions. And organisations need to be able to store original content, unalterable media, as well as new 'unstructured' file types including: memoranda, email, Instant Messaging, and other forms of digital information.

With many organisational processes moving from paper-based operations, compliance regulations require companies to demonstrate internal controls and processes in order to document what they do and how they do it, as well as demonstrate adherence to the regulations, so that in the event of an audit, they can show who had access to the data, when, and what actions were performed. A system failure or lack of visibility into the system is not an excuse for noncompliance.

Therefore, information must always be available for review by an auditor, with efficient accessibility of information and consistent availability – and this also requires the ability to produce reports that reflect origin of data and activity in real time.

1.2.1 What are the things organisations need to consider?

So the challenge for the IT manager remains: in the business world organisations need to deal with ever increasing volumes of information that are ever diverse, and increasing in size with every release of Microsoft Office. And, IT has to do this without extra budget or strain on the IT department's workforce, who, incidentally, are already working a 65-hour week. In addition, IT can't try to manage the storage resources by simply just adding more inefficient direct-attached storage devices, because that just doesn't work in the long term – and how do you successfully manage disparate storage devices anyway?

If storage growth is compounding at 50–100 % year, an organisation with one terabyte this year will potentially reach 32 terabytes in three years including backups. So not only have you got to put this stuff somewhere (more storage), but then you have to manage as well. Users still expect instant uninterrupted data; administrators face increased scalability and performance requirements, which are both initially unmanageable and invisible, with restricted (perhaps decreasing) budgets. In addition, board level executives need to ensure their company information is protected, accessible, and retained according to the latest worldwide, international, country and local regulations – regulations that are in their 10's of 1,000's and are constantly changing.

As a single example, in the USA alone there are over 10,000 US Federal Regulations surrounding electronic information retention. Extraction of data archive point-in-time views are becoming normal.

1.2.1.1 The problems

The main problems for an organisation are as follows:

- static or decreasing IT storage management budgets;
- multi-platform skills shortage;
- fewer IT system Admin Engineers;
- more sites, more data, more systems – all needing management;
- unstoppable data volume growth;
- globalisation – organisations now need to be available $24 \times 7 \times$ forever;
- compressed to zero backup windows;
- increased regulatory legislation around data management, IT and corporate governance;
- new communication types that need some sort of business policies set against them in risk mitigation
- inability to manage or control storage costs.

1.2.1.2 Things to consider

The main things an organisation needs to consider are as follows:

- What types of data does your organisation hold?
- Which of these data types need to be held?
- For what length of time does this data need to be held?
- Is any of this data likely to be used in the future?
- How critical is the data to the business?
- Who needs to access it?
- How quickly do they need to access it?
- Does it need to be held and produced in its original state (WORM)?
- If required, could you deliver every single instance of one type of specific data required by government legislation?

1.2.2 What does data lifecycle management mean?

1.2.2.1 What is IT Lifecycle Management? (Defining DLM/ILM/TLM)

Data Lifecycle Management (DLM) is a policy-based approach to managing the flow of an information system's data throughout its lifecycle – from creation and initial storage, to the time it becomes obsolete and is deleted, or is forced to be deleted through legislation.

DLM products attempt to automate processes involved, typically organising data into separate tiers according to specified policies, and automating data migration from one tier to another based on those criteria. As a rule DLM stores newer data, and data that must be accessed more frequently, on faster, but more expensive storage media. Less critical data is stored on cheaper, but slower media.

Early types of DLM tools included HSM. The hierarchy represents different types of storage media, such as RAID (redundant array of independent disks) systems, optical storage, or tape, each type representing a different level of cost and speed of retrieval when access is needed. Using an HSM product, an administrator can establish and make policies for how often different kinds of files are to be copied to a backup storage device. Once the guideline is established, the HSM software manages everything automatically. Typically, HSM applications migrate data based on the length of time elapsed since it was last accessed, whereas DLM applications enable policies based on more complex criteria.

The terms Data Lifecycle Management (DLM), Information Lifecycle Management (ILM) and Total Lifecycle Management (TLM) are sometimes used interchangeably. However, a distinction can be made between the three.

- DLM products deal with general file attributes, such as file type, size, and age.
- ILM products have more complex capabilities. For example, a DLM product allows searching of stored data for a certain file type of a certain age, whereas an ILM product allows searching of various types of stored files for instances of a specific piece of data, such as a customer number.
- TLM products allow formulating complex requests across multiple storage tiers and heterogeneous operating systems to

provide a more complete approach to managing all structured and unstructured data.

Data management has become increasingly important as businesses face compliance consequent to modern legislation, such as Basel II and the Sarbanes-Oxley Act, which regulate how organisations must deal with particular types of data. Data management experts stress that DLM is not simply a product, but a comprehensive approach to managing organisational data, involving procedures and practices as well as applications. Fundamentally what has happened over the last 15 years, since the advent of 'Open Systems', is that the ability to process information in a coherent, cohesive and consistent manner has been lost, or at the very least, seriously mislaid.

It would be quite a powerful technology that could examine an organisation's data storage and re-file all data consistently, in an intelligent manner, and that would allow the organisation not just to retrieve information easily (because every file system would be standard), but to store that data logically together in appropriate batches with like-times for deletion, as well as migrating data to upgrade storage – keeping the integrity of the data intact – and bringing a copy of the appropriate version software with it so it can be read in the future.

Suppose it is important to find data in the future and that it is not conveniently located where one would expect to find it . . . necessarily. So what? What drives the need for DLM or ILM products and services?

• Emerging regulatory and compliance issues (Data Protection, HIPAA, International Accounting Standards, Sarbanes-Oxley, Basel II, etc.), which drives

 – unbridled data growth (both structured and unstructured data), which promotes

 – the variability in value of data that an organisation owns.

• Organisations continue to pressure CIOs to manage more with less, and to control costs, so

 – it is becoming increasingly difficult, nay, down right impossible, to manage an organisation's data manually across an increasingly distributed and complex environment with any kind of hope of success.

This has not always been the case. Back in the 1970s and 1980s, mainframes kept all the data in logical file systems as previously mentioned. However, since the arrival of Client Server/Open Systems in the early 1990s, the art of information management has been lost. Basically, personal record keeping has become a chaotic free-for-all. Each individual stores and saves his or her data in different ways. According to a leading analyst, 60 % of all data is unstructured – our email, file and print servers (word docs, XLS spreadsheets etc.). How then, does one find a specific piece of data from an employee who worked at the company for four years and left two years ago? With no 'due process' it's not easy – and there have been several organisations that have been billed in the six figure region just to find and retrieve the data.

To make matters worse, all offices started to go 'paper-free' around the early 1990s. Prior to this, all organisations had to store their information in hard copy storage systems, including Microfiche, all of which were fairly sophisticated with offsite, fire-proof, storage facilities and processes behind filing and record keeping as well as audit trails to show due diligence. However, since the early 1990s these hard copy storage warehouses have slowly but surely disappeared, replaced with electronic data ware-houses. Paper records have disappeared and have been replaced with electronic data. To make things even worse, organisations now have additional communication methods and a range of elec-tronic routes to market.

1.2.3 Why is IT lifecycle management important?

It should be obvious that organisations must manage and store data more effectively. The upside is that ILM/DLM makes good business sense: in fact, that's why it existed in the first place. ILM/DLM is a prerequisite for good corporate governance, but is also an integral part of good business conduct. It protects repu-tations and manages risk, as well as promoting a safe, secured transaction environment. It protects global financial market safety and stability as well as tracking suspicious customers' movement. It adds value to customers' confidence and with it competitive advantage. It helps prevent terrorist money-laundering activi-ties and harmonises international regulatory approaches. Why wouldn't anyone want to know about Data Lifecycle Management?

1.2.4 Goals of data lifecycle management

Data is one of the most important organisational assets.

The above statement must be pure plagiarism. How many books, white papers, web sites or articles have made that statement? How many analysts, journalists, sales managers, business managers, business gurus, marketing managers, operational managers, database administrators and system administrators and have been bleating on about the benefits of looking after an organisation's data and information? Surely businesses must have caught on by now? Possibly, but unfortunately, probably not; but the scene is changing. Instant data gratification is out and data longevity is in. With an increasingly compliant and litigious society, data must be kept and accessed for longer.

Data as an asset is important in providing organisations with valuable information. Data becomes information and information becomes knowledge. This book discusses the differences between Data Lifecycle Management, Information Lifecycle Management, and Total Lifecycle Management in detail and examines the dichotomy at length. Although the principles behind the three concepts remain fundamentally different, it is all still data. Information management suggests that an organisation has done something intelligent with its data, and knowledge suggests that some cognitive process has been applied to that information.

From a technological point of view it is easy just to refer to DLM, and so initially we need to describe the fundamental goals of Data Lifecycle Management and its platform.

- **To make an organisation's data accessible.** All data should be readily available to support the businesses to which it is purposed. Availability requirements should not be restrictive.

- **To have an adaptable design and architecture.** Data continually changes. Hence, the processes, methodologies and underlying technologies that manage it should adapt to meet growing data demands.

- **To provide operational security to the asset.** The data management platform coupled with its process and methodology should provide auditing, tracking, and controlling mechanisms to manage the data effectively. Specifically, it must provide a complete management infrastructure that affords greater visibility into its daily use.

1.2.4.1 What are the technology trends we will see over the next few years?

Here are some of the expected technology trends:

- **Hierarchal Storage Management** (1997–2005). Vendors are already veering from using the HSM term and using both DLM and ILM instead. Although the term ILM is fraught with confusion and conflicting interpretations from various vendors (depending what technology they offer), vendors are already introducing numerous technologies that will be the starting point for the development of automated ILM products.

- **Data Lifecycle Management** (2003–2006). In the last few years, DLM products have emerged – the father of ILM if you like. With the increase in retrieval requirements through compliance issues and other emerging regulations, organisations have started exploring new ways to backup, store, manage and track their most critical data starting with virtual tape and disk-based backup. They have also started to implement some basic tiered storage capabilities, such as moving 'stale' data from their high-performance disk arrays to more cost-effective systems or deleting irrelevant data (MP3 files) altogether.

- **Manual Information Lifecycle Management** (2004–2007). This technology has the capability to index, migrate and retrieve data as well as prove its authenticity – on any part of the infrastructure. Although still a manual process when setting policies against business requirements, this is the point where ILM gives organisations the ability and technological intelligence to implement more-powerful storage policies. Some organisations will utilise virtualisation applications which logically group many arrays into a single 'virtual' storage pool and host them on emerging 'smart' storage switches. Manual ILM will provide users with a single logical file system view that is, in reality, scattered across multiple media types in multiple locations. ILM will enable companies to move data fluidly within the storage infrastructure as their evolving policies dictate, while shielding administrators and users from the underlying complexity.

- **Automated Information Lifecycle Management** (2006–2008). Automated ILM will integrate products that manage storage, virtualisation, and the data itself. Numerous storage management stack aspects will be imbedded into the infrastructure.

Compliance capabilities (retention, deletion, etc.) will also be imbedded into a number of storage management products and be well established in some vertical markets, especially the financial sector. The transition from manual ILM to automated ILM will require additional technologies in order to manage data as 'information', as opposed to managing it as 'data'.

- **Automated Total Lifecycle Management** (2007–2010). The total cost and value of a piece or set of data depends on every phase of its lifecycle, as well as on the business and IT environments in which it exists. TLM will automate the way that organisations look at their entire data set. TLM will offer organisations the ability to protect against media obsolescence, legacy data, future hardware changes, as well as dealing with all manner of diverse mobile assets, automatically managing storage costs and data movement (to lower cost storage options) as required, providing audits where required without the need for manual intervention. In other words all an organisation needs to do is decide on the data policy and TLM does the rest.

2
The Changing IT Imperative

Enduring change manifests itself in all societal, life and technology dimensions. Today, as companies continue harnessing the Internet's expanding power and reach, they not only exploit the Internet's potential to provide global portals into computing infrastructures for goods and services, they use it to access their own internal operations. Clearly, the demise of the explosive *dot com* era did not herald the end of electronic commerce. In fact, business-to-business commerce growth has steadfastly continued despite well-documented historical failures.

As a result, companies have now accumulated, and continue to accumulate, vast amounts of critical data that they store in various ways. Typically, these storage resources comprise dissimilar storage devices that different software programs reference using different methodologies and assumptions. These various hardware and software elements are collectively referred to as *heterogeneous storage*. And, as companies continue accumulating increasing amounts of data, these heterogeneous storage resources must necessarily accommodate the data expansion. Consequently, to provide enterprise applications storage, companies need to adapt, add, monitor, report and manage additional storage quickly, efficiently and seamlessly. Here, we are therefore discussing intelligent storage infrastructures that mere mortals can administer.

Data Lifecycles: Managing Data for Strategic Advantage   Roger Reid, Gareth Fraser-King and W. David Schwaderer   © 2007 VERITAS Software Corporation. All rights reserved.

Flexibility is therefore a key ingredient within solutions that help companies achieve their data storage objectives. When referring to the term *flexibility*, consider an architecture that is imminently responsive to change – in other words *adaptable*. Now, having accumulated oceanic amounts of data, the consuming burden that comprehensive data management subsequently presents, combined with today's litigious societal demands, highlights the importance of providing an extensible, resilient framework to manage data resources. It follows that data's continuing explosive growth is indeed the driver for the latest way-station in data management's evolution – *Data Lifecycle Management* (DLM).

Noting that a given storage device may well be inexpensive, it is imperative to recognise the significant costs beyond its initial acquisition. Storage management (including maintenance), electrical power, facilities floor space, etc., all quickly comprise *significant* follow-on expenses. Moreover, the typical absence of general data organisation heterogeneous storage further exacerbates data management challenges, leading to inefficient storage capacity utilisation *at best*. It follows that initial storage cost is often not the most important cost consideration over time. And, not only is there always more data to manage, but also the criticality of almost all data continues increasing which amplifies concomitant data protection and availability needs (read follow-on costs).

Lastly, recent regulations and stricter enforcement of antecedent regulations are now forcing companies to retain more data, for longer periods, and in *very specific* ways. For most enterprises, heterogeneous storage's general absence of data organisation further complicates this innately difficult endeavour.

Effectively addressing this spectrum of challenges often requires a spectrum of available technologies and methodologies. However, this approach further exacerbates data management challenges when different vendors provide the selected technologies. The usual consequence of this decision path is a pronounced inability to share hardware resources to reduce costs where applicable (read chronic, unnecessarily high, associated costs).

The following summary describes key factors driving contemporary intelligent storage platforms and associated DLM methodology developments:

• uncontrolled data management costs;
• unorganised and undiscoverable data;

- inefficient storage utilisation;
- emerging regulatory and compliance measures (DOD, HIPPA, Sarbanes-Oxley, etc.).

Presently, the storage industry references significant amounts of vague terminology. Therefore, this chapter introduces an overview of the difference between *Information Lifecycle Management* and *Data Lifecycle Management*. Subsequent chapters continue the exploration of this subject. Although it is impossible to resolve all terminology disputes within these initial pages, this book attempts to adhere to a very specific set of meanings.

To address today's data management challenges, a DLM methodology is needed. This fact is presently transforming the IT industry as we historically know it. DLM is not only a philosophy and process that leverages technology, but also enables additional enterprise processes to manage data effectively. These collective processes vary greatly and are usually unique to any given organisation. Therefore, consider the following generalised list of their activities:

- capturing and communicating current business processes;
- visualising, specifying and documenting current business requirements;
- specifying, visualising and constructing valued intelligent technology.

Given those boundaries, a solid foundational storage architecture coupled with a DLM methodology also suggests a clear requirement to provide enterprises with proper business forecasting and management tools. Providing such tools positions enterprises for the impending and substantial data management changes that enable enterprises to address the challenges discussed above.

Although implementing DLM solutions can prove a daunting task, those involved must also excel at balancing the methodology and enterprise alignments. Typically, organisations use technologies to disseminate information throughout the company to solve certain business problems. However, technologies are only tools and without a clear understanding of how to instantiate technology potential within enterprise focus, tools invariably prove as useless as they are expensive.

With that in mind, the salient concept is that of an intelligent data management platform coupled with a methodology. However, it should be readily apparent that the data management platform should accommodate heterogeneous operating systems and storage devices commonly endemic in most data centre environments. This consideration provides the above discussed flexibility by providing an ability to add storage devices quickly to a shared mass storage pool. Moreover, companies having the tools to report and manage the storage resources in such heterogeneous environments have the ability to increase a company's return on investment, and thus possess the potential to decrease naturally growing infrastructure costs while simultaneously improving business processes.

In essence, such a data management platform would comprise an enabler that could rapidly address specific critical business problems. Consequently, most companies have compelling business reasons to implement such a platform. After all, organisations often share a common characteristic: an organisational sense of urgency caused by compelling business motivation. Here, compelling motivations such as agile competition and rapidly changing competitive landscapes usually prove enough to warrant platform implementations.

In other words, if you or your executive staff have determined that enterprise survival over the next decade depends upon having the ability to manage and control storage costs, few are going to question a data management initiative that envisions an entirely new way of conducting business.

In the final analysis, if you cannot manage enterprise storage resources and data, all attempts to place value on that resource are futile. More importantly, if a company cannot manage the growing and *living* data, attempts to meet regulatory compliance measures will place burdens upon the entire business infrastructure in order to reduce the likelihood of incurring monetary penalties.

2.1 Introduction to utility computing

Utility computing is the IT industry's latest computing model. The intention is to provide IT services for geographically dispersed users via highly effective and efficient use of shared resources. Before proceeding, note that because of its complexity this discussion does not attempt to cover all aspects of utility computing. However, a quick overview of how it affects the IT industry and how it relates to data protection tools and solutions and methodologies such as DLM is useful. Thus, this brief introduction provides an opportunity to see how this powerful model can provide a generic IT service framework.

By itself, utility computing is a vital and integral component in the overall enterprise data protection arena. Moreover, it additionally accommodates DLM because it naturally addresses compliance requirements through its management of data from creation to demise. In addition, it not only enables many enterprises to meet regulatory compliance measures, but also enables the enterprises to reduce costs by moving data to lower-cost media via *tiered storage architecture* approaches. Using this computing model also provides organisations with chargeback mechanisms to reduce the cost centre burden. Figure 2.1 provides an illustration of a typical service model that most closely aligns business processes with IT infrastructure.

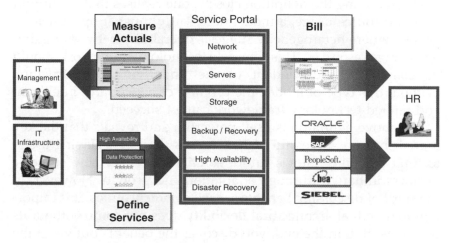

Figure 2.1 Basic elements of a typical service model

By giving IT service departments the ability to create chargeback mechanisms to lines of businesses and delivering reporting capabilities to the lines of business, utility computing provides substantial advantages for an entire enterprise. It does this because collaboration and communication between the two different groups produces a company coherency that is more competitive and powerful than possible when the two entities operated independently in the presence of an informational gap. Therefore, the earlier discussed flexibility requirement that utility computing naturally delivers can tightly couple previously separate and distinct organisational units.

That said, the most challenging IT industry undertaking is to define utility computing clearly. There have been many attempts to define utility computing successfully under different names by companies such as Symantec, IBM, HP and Sun. Perhaps the most are familiar is IBM's 'On Demand' strategy. Therefore, before continuing, consider the following utility computing definition which clearly describes the utility computing model in a very efficient and elegant manner:

> *The IT utility provides a fast, reliable service to end users, and can adapt to changing business needs without human intervention. At the same time, the IT utility makes use of shared hardware resources, which maximises utilisation and reduces overall costs.*

By examining the definition closely, one realises that to support shared resource usage, a model necessarily must support heterogeneity where heterogeneity suggests operability between dissimilar storage resources. This complete technology framework should be completely independent of the underlying infrastructure. Without this type of model, the promise of a standard, intelligent and unified technology framework cannot succeed.

The above definition is itself powerful, and reveals that the key ingredient to a successful architecture is *flexibility*. Having architectural flexibility allows future solutions to meet new strategic business initiatives through adaptation and growth. However, in the spirit of discussing benefits as well as practicalities, it is important to note that architectural flexibility does not come without its problems. But, in the end, you discover the benefits outweigh the initial concerns involved in integrating and building a heterogeneous infrastructure.

An example of a heterogeneous infrastructure would be unique virtualisation applications which give the ability to share data serially among heterogeneous systems that access the same storage devices. Another example would be file systems that provide the ability to align the file system across differing classes of storage devices such as inexpensive disks or high-end storage arrays.

In addition, tools to automate the classification of data comprise a final example. These types of capabilities, coupled with strategic data classification framework and methodology, help enterprises to bridge the gap between managing complex distributed storage infrastructures, reducing costs and complying with emerging regulatory compliance laws.

2.2 General market highlights

Earlier discussion described the changing imperative of IT, including a utility computing primer and how it relates to data management. However, because a new and genuine issue is that directors or board of trustees now have a legal obligation and a duty to act with reasonable skill, care and diligence to safeguard the shareholder's or company assets, it is useful to provide general market statistics on compliance and its associated effect on information management as well as the entire storage industry.

Suppose that a matter previously resolved with normal operational activity has surfaced and threatens the way an enterprise conducts business. Compliance legislation mandates that many records will need to be retained for various periods of time. But simply retaining information is not the point; retrieving the information easily and in the most efficient manner is now paramount. Therefore, this review examines data growth across various market sectors. This reveals the parasitic growth rates among these various sectors and the effects that it can and does place on companies around the globe.

That said, it should be apparent that the growing need to capture and retain data continues increasing at alarming rates, special thanks to compliance and data conservation laws not withstanding. With earlier discussed factors including the Internet, e-business, e-business and the growing global economy, it is clear that the world is getting smaller and more complicated. Moreover, there is also a renewed expectation that companies will protect profitability and shareholder value.

These factors make it increasingly urgent for one to understand a company's data and what is necessary to optimise the management of its critical storage resources. This need for a better and quicker understanding of the value of data has spurred rapid industry growth in the data storage management market as shown in increased product development, establishing it as a major category of critical business utility software. However, despite the fact that abundant storage products are now available, the challenges remain for companies to understand how to achieve improved efficiency and profitability by aligning available software solutions horizontally across the various

- departments;
- divisions;
- geographic locations;
- applications;
- lines of businesses;
- processes.

A comprehensive solution requires aligning business priorities with IT and providing a data lifecycle management strategy that includes automated federated policies, procedures, and technologies to track the data. Incorporating all these factors and aligning them across each dimension of the IT infrastructure is the ultimate goal.

2.2.1 Current storage growth

For the past several years, enterprise data storage has traditionally involved system administrators and managers below the executive ranks. This typical stratification essentially divided the corporation into separate domains, preventing the development of a unified strategic process and strategy for bridging the inevitable information gap between business groups and IT. Even though many storage professionals have the skills and knowledge to implement and manage storage solutions, it is likely that few are aware of the various business drivers such as return on investment and total cost of ownership.

Furthermore, with storage now accounting for approximately of 40–60 % of the average IT budget, there is still general agreement that most IT organisations do not have a cohesive management strategy regarding storage or data retention as relates to emerging compliance issues. However, industry movements in the past year have responded to this problem, enabling companies to leverage business intelligence into their enterprise storage solutions to produce positive business results. Now, storage management solutions are becoming storage *and* executive information systems, simultaneously serving executives, front line managers and analysts as well as storage industry professionals. Furthermore, this intelligent infrastructure typically provides the ability to meet emerging and current regulatory archiving demands.

In this section we address these particular developments by carefully examining and outlining the following points:

• the largest capacity of compliance records;
• the fastest growing media type;
• the fastest growing market sector.

In order to provide an idea of how existing and emerging compliance laws affect storage, we discuss several examples. However, before reviewing the statistics, it is necessary to describe some terminology first. Therefore, in this context a compliance record is described as *those business record activities that are maintained in compliance with a particular law or regulation within a specific domain.*

Note that the illustration in Figure 2.2 is worldwide in scope and includes all vertical markets such as corporate, commercial

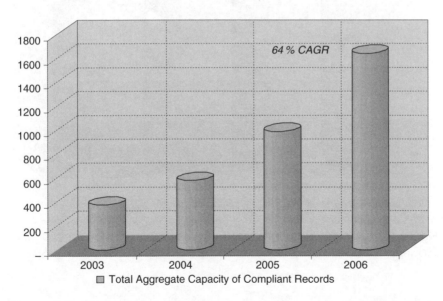

Figure 2.2 Forecasted compliant records capacity growth in Peta Bytes (Source: Courtesy of Enterprise Storage Group 2003)

and government sectors. It clearly identifies the total aggregate capacity of compliance records.

The total capacity of compliant records will increase from 376 TeraBytes (TB) in 2003 to over 1 PetaBytes (PB) in 2006, representing a CAGR of 64 %. And, as storage growth rates continue to increase across the board one would assume that the compound annual growth rate will continue to grow as well.

Figure 2.3 provides a graphical representation and view of the different storage media storing compliance records. It reveals that disk-based media continues to grow steadily. Furthermore, the graph depicts the forecast of compliant records capacity growth in PetaBytes and further categorises them by media type. Moreover, it continues to highlight the increase in compliant records created and stored in conjunction with the need to manage the ever-increasing volumes of data and its haste to shift from offline to online digitised records management.

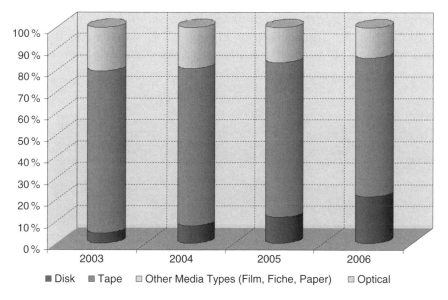

Figure 2.3 Forecasted compliant records capacity by media type (Source: Courtesy of Enterprise Storage Group 2003)

Figure 2.3 clearly indicates that tape continues to be the most prominent storage type on the market.

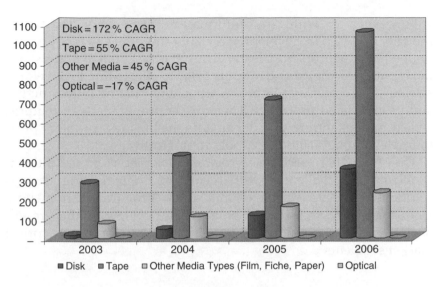

Figure 2.4 Comparison of different media types (Source: Courtesy of Enterprise Storage Group 2003)

To further illustrate this point, Figure 2.4 provides a comparison between tape, disk, optical and other types of media. Understandably, after reviewing this graph one can certainly see that the demise of tape might well not be as imminent as one might have thought.

Figure 2.4 indicates that disk-based storage solutions are the fastest growing media segment. It further predicts that the worldwide capacity of compliant records stored on disk-based solutions will increase from 17 PB in 2003 to 353 PB in 2006, giving disk-based storage solutions a 172 % compound annual growth rate. On the other hand, tape-based storage solutions represent the most prevalent media segment.

Later discussion will show that tiered storage architectures that exploit newer disk technologies will continue to grow. However, tape based solutions will also continue to grow, albeit at a lower recognised rate. Statistics such as this support the idea that enterprises require complete data management solutions that support growth. So, a company truly benefits from developing global strategic policies and a global data lifecycle management technology infrastructure to assist with the explosive growth of living data.

However, as with any information technology initiatives, the very first step in developing a viable strategy is to identify and understand the problems clearly. Thus, it is essential to understand the inputs, processing and outputs prior to attempting to develop a strategy that corrects or improves upon any existing infrastructure and its associated processes. Therefore, a proper strategy requires a clear understanding that documents the business problem prior to implementing technology tools strategically.

The following sections provide discussion on additional problems and challenges with regard to implementing various solutions to help manage this data growth phenomenon. Understandably, this should be a general review of the growth of data, but also provide additional data related to a few critical market segments.

2.2.2   Enterprises for which DLM is critical

Because recent compliance laws have placed huge demands on managing the lifecycle of various types of data, DLM solutions could not have emerged at a better time. These regulatory requirements require enterprises to identify, restructure, move and archive unstructured and structured information for a varying amount of time with the additional requirement of being required to retrieve it on demand. Figure 2.5 provides a strong sense of the number of compliance records by various industry sectors.

Figure 2.5, indicates that Life Sciences is the industry with the fastest growing capacity of forecasted compliant records, easily verified by quickly reviewing these growth patterns. Equally important, the worldwide capacity of compliant records in the Life Sciences industry will increase from 30 PB of data in 2003 to 192 PB of data in 2006 – a compound annual growth rate exceeding 86 %.

This is graphically illustrated in Figure 2.6, which indicates that the Life Sciences sector will need an intelligent data management infrastructure that possesses the ability to meet regulatory compliance legislation demands.

Figure 2.7 provides a more detailed summary of this market sector by illustrating the forecasted storage media type. The graph indicates that disk-based storage is the fastest growing media type, representing a compound annual growth rate (CAGR) of 67 %

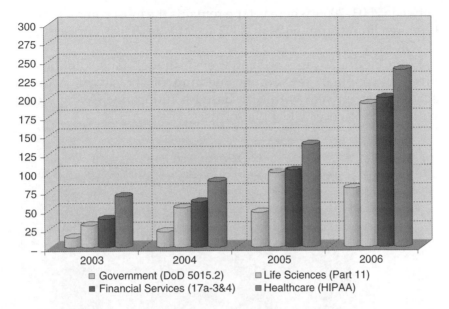

Figure 2.5 Forecasted compliant records capacity growth in PetaBytes by various industries (Source: Courtesy of the Enterprise Storage Group © 2003. Printed with permission)

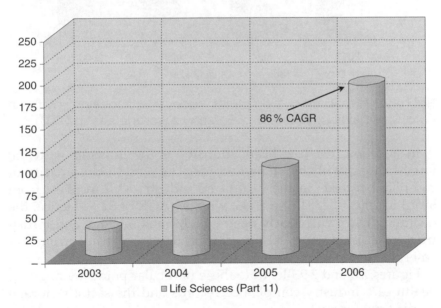

Figure 2.6 Forecasted compliant records capacity growth in PetaBytes (Source: Courtesy of the Enterprise Storage Group 2003)

between 2003–2006, whereas tape remains the most common media type, declining from 66 % in 2003 to 50 % in 2006, representing a CAGR of −9 %.

From Figure 2.7, it is clear that even though tape solutions are presently popular, the storage medium of choice will begin to change year on year with disk-based solutions increasingly used within Life Sciences DLM platforms.

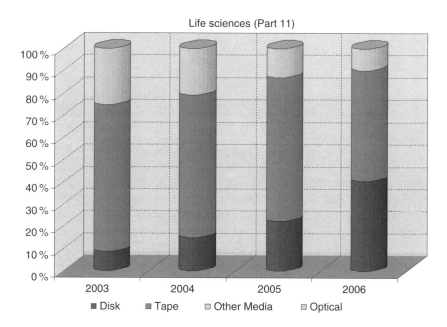

Figure 2.7 Forecasted compliant records capacity growth in PetaBytes (Source: Courtesy of Enterprise Storage Group 2003)

In addition, the healthcare industry is the industry with the largest capacity of compliant records. The worldwide capacity of compliant records in the healthcare industry will increase from 68 PB in 2003 to 238 PB in 2006, with a CAGR of 52 %. Hence, the health industry will clearly need a DLM methodology and infrastructure assistance to meet the growing demands from current and emerging compliance laws.

Figures 2.8 and 2.9 illustrate these particular points by aligning health care industry compliance records and the sector's storage media type forecast.

Figure 2.9 indicates disk-based storage is the fastest_growing media type in healthcare, representing a 170 % CAGR between

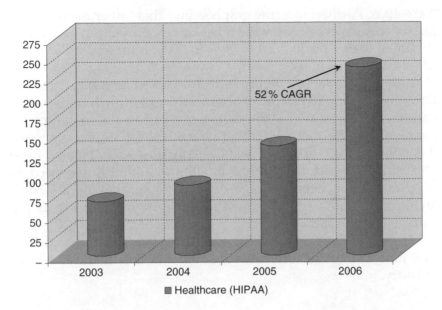

Figure 2.8 Forecasted compliant records capacity growth in PetaBytes (Source: Courtesy of Enterprise Storage Group 2003)

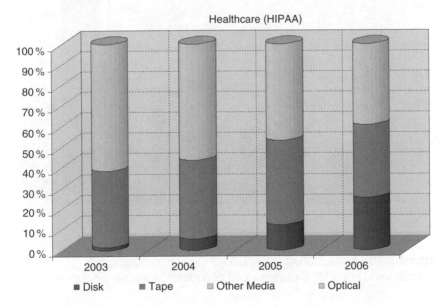

Figure 2.9 Forecasted compliant records capacity growth in PetaBytes (Source: Courtesy of Enterprise Storage Group 2003)

2003–2006. Furthermore, the graph shows that 'other media' (film, microfiche, paper) remains the most common media type, despite declining from 61 % in 2003 to 39 % in 2006, representing a CAGR of −14 % for this market sector.

Although the above figures are forecasted graphical illustrations, they clearly depict the enormous storage growth rates among various industries and stand as a testament to this data growth phenomenon. Furthermore, due to the advanced digital economy, it is unlikely the market will stabilise any time soon. Finally, these illustrations depict that future intelligent architectures need to support heterogeneous storage devices. Again, having a flexible architecture is the key in providing for growing and expanding storage infrastructure.

Figure 2.10 illustrates another way to express this phenomenon, this time using the forecasted growth for the financial service industry.

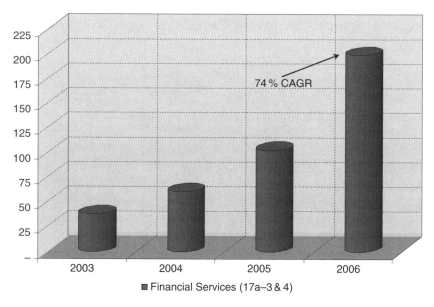

Figure 2.10 Forecasted compliant records capacity growth in PetaBytes (Source: Courtesy of Enterprise Storage Group 2003)

Figure 2.11 confirms this financial sector view using another media forecast summary. It is truly startling to see that the capacity

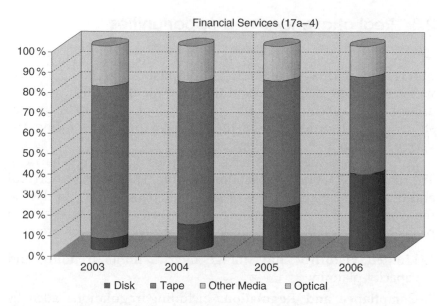

Figure 2.11 Forecasted compliant records capacity growth in PetaBytes (Source: Enterprise Storage Group 2003. Printed with permission)

of compliant records industry will grow from 38 PB in 2003 to 200 PB in 2006, representing a 74 % CAGR.

At first glance, the four bar graphs in Figure 2.11 might appear to be identical, but they are not. Carefully looking at Figure 2.11 reveals that disk-based storage is the fastest growing media type, representing an 89 % CAGR between 2003–2006 and that tape remains the most common media type, declining from 74 % in 2003 to 48 % in 2006, representing a CAGR of −14 % for this market sector.

The preceding discussion clearly underscores the growth of data and the overall lasting effects compliance would bring. In summary, it is clear that

1. Data retention is growing in all segments.
2. Disk is growing as a fraction of all digital data storage.
3. Compliance comprises an enormous storage management factor, requiring even more data retention.
4. Enterprises must manage data intelligently through its lifetime because there is enormous data to store and numerous media choices.

2.3   Real challenges and opportunities

In the previous sections, we reviewed general storage growth trends and reviewed the effects compliance records have on various market sectors. In addition, we identified the need for a clear strategy coupled with adequate tools to help with that strategy. This section covers the different problems facing the various information technology sectors.

Data access issues experience numerous technology barriers. The catalyst for action is most often tangible. Therefore, resourceful and disciplined management of key challenges is the vital key in the overall business strategy. You might ask: 'What are the real problems we face today?' Some particular problem areas are:

1. **Limited visibility** into storage metrics limits trending and capacity planning.
2. **Compliance and Regulation,** including regulations such as Sarbanes-Oxley Act of 2002 and the Health Insurance Portability and Accounting Act (HIPA) of 1996, and auditing.
3. **Managing heterogeneous environments** addressing different problems and solutions.
4. **Excessive storage total cost of ownership** (TCO) attributed to reactive management.
5. **Phenomenal data growth**.

Although the preceding list is not exhaustive, it does suggest some of the salient problems facing IT. And, the enumerated items apply to various market segments *across the board* ranging from financial services, life sciences, healthcare and government entities to name a few. Just one example, is the previous discussion, which revealed the effects that compliance issues cause in data storage consumption.

2.3.1   Real issues identified

As we proceed into the new millennium, we will see an evolution of the typical trends business enterprise computing infrastructures have experienced. Previous decades witnessed numerous data storage management technology advancements; this decade is proving no different. It has been said that the Internet is a *de rigueur* component of most business networking strategies that provides ubiquitous access to corporate applications and services both to internal employees and corporate customers worldwide. And, as you have seen in the past several years the Internet has been utilised immensely with new business to business applications and the like.

So, what are the challenges data management solutions present? Although there are no quick and easy answers, the experiences of successful organisations reveal that *extensive and meaningful managerial and executive storage applications are the key ingredient of an intelligent infrastructure*. Providing key business managers and storage professionals with tools that assist their decision making process is truly the next storage technology imperative. Moreover, experience also indicates that these solutions must support heterogeneous systems. An enterprise may desire a single operating and storage sub-system infrastructure, but the reality is that this is an unrealistic aspiration and the solution is an agile, virtual, layered solution that can adapt and grow as business continues to evolve.

2.3.2 Data compliance

As time passes, the number of regulatory agencies and their associated regulations will continue expanding as the previous discussion indicated. That discussion captured the essence of what 'growing' data really means. Therefore, corporations must be prepared for and respond to increasingly complex, and even contradictory, regulatory audits. To do so effectively – without diverting precious IT resources – companies will also need a compliance solution that is fast, flexible, fully automated, and integrated with existing storage and data protection architectures.

Here are a few examples of genuine, pervasive storage industry issues:

- Regulatory requirements are driving data growth and mandate longer retention.
- New file types such as unstructured data (including email, memoranda and instant messaging technologies) must now be regarded as mission critical information.

- Global business forces data availability and 24×7 access across multiple time zones with planned outages increasingly viewed as unacceptable.
- Auditing systems must be readily available.

2.3.2.1 In many ways, compliance data management is simply a specialised form of data management

The material this chapter discusses, as well as the remainder of this book, is presented in the context of storage and data lifecycles referred as Data Lifecycle Management or DLM. Often, the more specific context of data and managing its data lifecycle is used for demonstrations, examples and explanation of the various concepts involved.

This DLM context comprises the arena in which storage management and the use of data analysis and management has occurred and continues to occur. Thus, the material generally considered problems where the various entities under examination are *data* and the variables of interest are *attributes* of the various different lifecycles of this living data.

Keep in mind that general DLM is valuable and applicable to a wide range of business entities. As electronic communication continues, more emphasis will be placed on managing the tremendous amount of data that continues to grow despite a sluggish economy. Thus, though storage management products have been in place in different forms in the data centre, DLM acts as a component of the overall utility computing paradigm. Data Lifecycle Management clearly illustrates that having the ability to manage an operational management asset such as data effectively clearly provides a competitive advantage in the overall information technology lifecycle. By mapping business strategies and initiatives to the various business technology requirements, businesses have the ability to gain a competitive advantage over their competition.

Decisions and architectures are made from information; however, a decision is only as good as the information it is based upon. Therefore, as mentioned earlier, we will provide real life examples of problems that companies are facing. With that said, let's discuss at a high level the importance of how an effective active storage reporting management platform can assist an overall DLM strategy.

2.3.3 Case study in ineffective storage reporting

In a storage management assessment project for a large financial institution, the question emerged: *Where does storage reporting information originate?* In answering this question, we found that the information came from a myriad of sources. However, these sources were various structures and systems that different outsourcing companies developed to keep the company appraised of ongoing events such as current storage utilisation. These crude reporting tools were created to provide guidance information regarding their total storage environment. Moreover, processes were in place to process the flow of information on numerous aspects of the enterprise's storage infrastructure.

Although these reporting tools served major business functions within the company, they also exacerbated internal support weaknesses. For example, many managers were seeking information pertaining to their storage environment. They typically sought the information quickly because their next higher management layer required the information to make an informed business decisions concerning additional storage hardware purchases. However, because of their design, the existing tools had an overall detrimental effect on centralised reporting and analysis because the tools did not provide for ease of access or a standard mechanism to report across a wide variety of platforms.

A brief assessment review revealed how incomplete logical data displays caused people to overlook critical pieces of a problem not addressed in the data analysis phase. The internally consistent logic masked the fact that the basic storage assumptions were faulty or incomplete, and that the data comprising them were questionable in the first place. Therefore, this process proved the importance of having proper storage resource management tools to properly process analysis on a given set of data. This is nearly always the precursor to pursuing a DLM strategy actively that involves the classification of data.

Why is that? Because if you cannot report on the data you currently have, how can you place a value on it? Moreover, implementing DLM tools will fail if you cannot report on the data's location, type or access rates. Ascertaining the relationship to the existing applications to determine the data's importance is crucial in determining appropriate service levels, or meeting regulatory compliance laws.

Even under the best circumstances, the existing reporting conduits examined in this project simply pointed out that a problem existed. The discrepancy in the numbers (often significant) forced independent judgment calls. Eventually, it was shown that the information flowing through ineffective storage tools could be skewed at any given moment without the proper tools to present the information properly. Unfortunately, this type of channelling also tends to distort valuable information that can help make informed business decisions. Therefore, effective storage resource management (SRM) tools have the potential to provide the foundation for data lifecycle management. Moreover, a centralised homogeneous platform provides businesses with a reporting layer for their existing storage environment.

During our discovery and assessment phase, it became apparent that the absence of a centralised reporting structure precluded obtaining quick and accurate data to make sound business decisions. As a result, it was noted that a more aggressive strategy with associated tools should be used to go outside normal channels to create checks and balances. This type of environment effectively leverages data and lifecycle management solutions.

In summary, given the dispersed and sometimes inaccurate data, we have given an example of how just using people as conduits without effective tools can lead to inaccuracies. However, by leveraging an intelligent storage platform coupled with an effective methodology, companies can keep managers and key personnel informed of the status of the overall storage infrastructure and at the same time provide them with an infrastructure that could reduce the administrative and user complexity in managing the day to day compliance and policies associated with various units of data. With a heavy accent being placed on storage administration via common federated policies with auditing capabilities, one would hope that it would be a valuable proposition to most companies.

2.4 Summary

Building a solid DLM strategy is fundamental to an overall storage management strategy. This strategy consists of not only a philosophy, but also a proven methodology and process that closely integrates a standard technology framework for managing data as an asset. In addition, it properly equips a business with tools to comply with existing and emerging compliance laws and regulations.

As the following chapters will reveal, a complete DLM infrastructure rests upon various technologies which compose an entire solution – one particular product does not make an entire DLM solution. You can certainly buy storage products to automate various facets of DLM, but it is necessary to combine them with a conscious decision to manage your data as a *business asset*, and an overall architecture based on the goals you wish to achieve.

Following this introduction, it is time to embark on this course of direction to clarify, describe and discuss the various ways to help manage the lifecycle of data within your realm. In doing so, it will also provide a great opportunity to see real-world software solutions and techniques that can be applied in your environment to assist in this management endeavour. This is an effort to help companies develop a foundational and storage management infrastructure – one which can give you a competitive advantage over the competition.

3
Being Compliant

Compliance is not a new phenomenon; it's actually been around a long time. Back in 1906, the International Electro-technical Commission (IEC) was established. In 1926, the International Federation of the National Standardising Associations (ISA) continued the IECs work, focusing on mechanical engineering. In 1946, delegates from 25 countries met in London to discuss international standardisation and as a consequence ISO (International Organisation for Standardisation) was formed and began operations on 23rd February 1947. Nowadays, ISO is a network of national standards institutes from 148 countries working in partnership with international organisations, governments, industries, business, and consumer representatives.

These various standards were originally designed for the manufacturing industry, but were later adapted into ISO 9000 for the services industry to provide customers with standardised products. And, any consumer purchasing a compliant product or service expects to receive a piece of good or service at a particular standard. If this standard is not met, then the provider is likely to lose customers and eventually cease trading.

If there were no standards, we would all soon notice. Standards make enormous contributions to most aspects of everybody's lives – although very often, that contribution is hidden. When there is a lack of standards their importance is brought home. Purchasers or users of products soon notice when the products turn out to be of poor quality, do not fit, are incompatible

Data Lifecycles: Managing Data for Strategic Advantage   Roger Reid, Gareth Fraser-King and W. David Schwaderer   © 2007 VERITAS Software Corporation. All rights reserved.

with existing equipment, are unreliable or, for that matter, dangerous. When products meet expectations, they are taken for granted. Most people are unaware of the role standards play in raising quality, safety, reliability, efficiency and interchangeability levels.

The idea of standardisation makes commercial and business sense. So, too, does legislation and policy, both of which provide protection for the consumer, but also protection for the genuine business against fraudulent and criminal activity and organised crime.

Basically, compliance has three different aspects

- **Standards** tend to lead the way in terms of providing a structure for legislation; they are regulations for business-to-business as well as business-to-consumer protection and include standards such as ISO17799, Basel II and the International Financial Reporting Standards.

- **Regulations** are country specific pieces of legislation, which look to improve business behaviour such as security, information protection and data recovery – these pieces of legislation can also have international consequences and include Sarbanes-Oxley, Data Protection Acts and Anti-Money Laundering Acts.

- **Policies** tend to be both governmental and business-related; they address operational risks and corporate governance control guidelines and include 'The European Commission' and 'Operational Risk Systems and Controls' guidelines.

So, if compliance has existed for 100 years or so, what is the fuss about?

Well, over the last 10 years, slowly but surely, businesses have achieved that Utopia that is known as the paperless office. As a consequence, audit trails have moved from paper-based processes into the data centre. Hence, existing legislation that audited any number of operations in paper environments is no longer appropriate. Over the last few years, new data compliance legislation, regional and international directives and quality codes of conduct have come into force to combat this problem. Furthermore, 15 to 20 years ago, most organisations carried out their business activity within a local region or were specific to a country. Now, companies conduct business internationally and therefore have more than local legislation to worry about. This affects all companies, large and small, and has become more confusing than similar legislation in the past.

By moving the audit trails away from the operation of the business into the data centre, the number of stakeholders involved with each audit trail or regulation has increased. As the business processes have moved into the data centre, we now have to include IT departments as well as the board, all of whom have a responsibility to ensure that audit trails are provable and support the business. 15 years ago it was so much simpler – a department that needed an audit trail could do it easily itself, setting standards and providing an audit trail that was accessible and transparent. Now, it is not so easy. To make matters worse, since September 2001, new investigatory and terrorism legislation has come into force which leaves IT departments having to pull proverbial rabbits from the hat – and then there's a bit of pressure because all organisations have to get this right to stay in business!

So what are the pain points facing organisations in the compliance heavy, litigious society we live in? What keeps executive up at night? Well, to start with, avoiding imprisonment springs to mind. But, fundamentally, organisation heads are governed first by the stakeholder and second by cost margin. Executives must ensure stakeholder trust as well as manage the costs associated with meeting regulatory requirements and minimising investment in redundant systems; there is also a concern that addressing compliance is distracting from revenue-producing activity. Compliance is much more than just installing software: *How do organisations bring together the business and IT side of the company to make sure that any effort or money thrown at any regulation will actually work?*

The bottom line is that compliance has been around for a long time and is here to stay. It is here to stay to protect investors and customers, to protect the business and employees and, in fact, makes good business sense since it plays an important role in any business's organisational processes. By getting compliant now, organisations can get ahead of the competition by using compliance standards, regulations and policies as essential tools to stay in business. Compliance isn't just about email, it's all about data. It is about the ability to readily deliver corporate data resources when required by law or regulation ... wouldn't it make sense to do this anyway, just for the sake of the business?

But why is Data Retention Important? Well apart from human error, application corruption, and hardware errors (disasters potentially ruining businesses and regularly costing thousands of Euros in lost business, as well as the potential loss of competitive advantage or the inability to do business), there are also issues such as huge risks to reputation where data is not secure and potential risks in damages for breach of contract – e.g. confidentiality agreements, etc. With the new compliance regulations, there is also a distinct risk of criminal liability – directors may be personally liable or risk disqualification – and the risk of organisations being sued when employees illegally hold unauthorised digital copies of music, images, software, etc. (which also wastes valuable storage space) is increasingly likely.

Compliance is not a *tick in the box* process – it is here to stay. A compliance strategy should be an ongoing process which provides organisations with a data protection and management strategy that encompasses all business areas on an ongoing basis. Chances are good that larger enterprise organisations are more than aware of the requirements for compliance. Certainly, financial services companies will have compliance officers, if not legal departments, who concentrate on compliance issues. However, for smaller businesses, as well as some of the smaller vertical markets, the raft of regulations, rules and policies represents a huge number of business and legal tripwires.

3.1   So what are the regulations?

Well, there are thousands of them and the number of regulations and their requirements are growing every day. It is simply not practical to enumerate all the regulations given that there is estimated to be between 10,000 to 15,000 compliance 'issues' in the USA alone and, indeed, probably as many, if not more, across the European, Middle Eastern, North African and Sub-Saharan (EMEA) regions. Instead we have chosen some of the more publicised and vertically aligned regulations as a simple example of what organisations are faced with.

In order to deal with such a range of constantly changing regulations, it is critical that any compliance investment is one that is flexible enough to grow with evolving compliance changes. And, for this reason, customers should also be wary of investing in *point* solutions. Giga Research analysts anticipate that companies implementing point solutions (i.e. email archiving) today will circle back in 12–18 months because they find their strategic objectives *were* met with point solutions. They quickly become unmanageable as each issue or application requires a new point solution to deal with each compliance issue. Before you know it, you could potentially have hundreds of point products each addressing a particular issue – but precious little IT governance!

Reacting to compliance, rather than planning for it, can be both resource intensive and financially intensive. Organisations need to implement an auditable, automated solution that ensures consistency of compliance operations. By being proactive rather than reactive to compliance, businesses will not have to constantly chase their tail but can ensure consistency of operations and stakeholder/shareholder trust, as well as good customer satisfaction and expectations across the business set and controlled operations.

So, there are *thousands* of regulations. It is not all about Sarbanes-Oxley, nor is it just about the data held in your email system – all records can be subject to regulation, not just email. Existing regulations are confusing and constantly evolving. They are not only country specific, they also apply to businesses trying to trade within that country. Following well publicised corporate scandals, the enforcement of these regulations is real and can be increasingly expensive. A few years ago organisations could simply say that the records had been deleted. However, today saying: 'But I don't have them anymore!' just doesn't work.

The financial implications of noncompliance are enormous. Over the last few years we have seen increasing fines levied against organisations who failed to produce data requested by, not only governmental, but also legal and financial bodies. The knock on effect, in these troubled times of, say, illegal money laundering, is taken very seriously . . . and not just by the FBI (who were criticised for the intelligence supplied prior to September 2001). The terrorist attacks in Spain and the UK have all meant that vigilance in this area is likely to increase. Because of various terrorism acts, organisations now have strict limitation periods placed on data and other forms of communications data, effectively escalating the amounts of data they need to backup and manage, for longer periods of time.

There are numerous international and European codes of practice regarding data retention and management, management of risk, corporate and IT governance, as well as horror stories relating to organisations' ability to deal with the issues. But what regulations affect all companies? Well, data protection and management is primarily the crux of all legislation. Organisations must have procedures for the fair processing and secure retention and recovery of personal data. Failure to do so results in an unnecessary loss of competitive market advantage as well as heavy potential fines.

Even a few years ago, legal departments advised operations to trash or shred old correspondence, which effected the removal of evidence of noncompliance. But since the Enron scandal, legal departments have done a complete reversal. This means that organisations must now alter their operation procedures and enforce document retention and lifecycle management into a process that has been a free for all for the last 10 years. This means that, for example, retrieving specific emails can be tortuous.

Sarbanes-Oxley (or SOX) affects all US Companies and Foreign US listed Companies. This means that for American companies, their subsidiaries, or even companies wanting to conduct business in the USA, this legislation imposes

- strict reporting and record keeping rules;
- commission rules and enforcement standards;
- annual accounting rules and standards;
- corporate governance controls.

These provisions force organisations to ensure they have auditing processes, quality control and corporate independence.

Regarding directors or those on company boards, companies must maintain an effective system of internal control to safeguard shareholders' investment and company assets, which can include disaster recovery and business continuity strategies and operational activities, as well as ensuring the organisation has an effective backup and restore process that is proven and can mitigate risk. Suffice to say, directors and those on company boards are liable for fines, disqualification and even imprisonment if they fail to safeguard company assets.

It is commercially prudent to retain data for use as evidence for the period in which a legal claim could be brought. Legal proceedings may be brought within the periods laid down in the various

Limitation Acts. Here, limitations are the length of time that data needs to be held before it can be deleted. Globally, these limitation acts vary from country to country and industry to industry, and even within individual industries. These limitations regulations scream out for policy based storage, data management, and near and offsite backup and archiving.

There are also copyright infringements and potentially damaging files held on corporate storage, servers and networks. In the last few years, an estimated 63,000 viruses have traversed the Internet, two-thirds of the music files on computers comes from illegal file swapping and two-thirds of corporate PCs carry some form of pornography on employee computers and the company network.

Today, using any readily available browser to connect with peer-to-peer software, it's very easy for people to download copyrighted material. Neat technology to be sure, but unfortunately, it can be illegal. Unauthorised digital copies/downloads of music, images, video clips, software, etc. may be held or distributed by employees whilst at work. These actions may infringe third party rights in those materials. This kind of noncompliance has been rife for years, but now organisations are stamping down on illegally held files, as well as becoming more vigilant when it comes to plagiarism, pornography, video piracy, etc.

3.2 Financial services companies

Data processing and data retention policy, a business critical issue in relation to the majority of organisations, assumes even greater importance in the context of financial services organisations, where investor trust and market and analyst perception are paramount. Even leaving aside the other likely consequences, the loss of reputation associated with the failure to maintain the security of investors' account details could be catastrophic.

Data processing and data retention is tightly regulated within the financial services sector. Hence, in addition to the requirements applicable to all organisations, there are numerous statutes, codes and guidelines covering data management issues, which specifically impact financial services organisations.

Regulations, such as Basel II, the International Accounting Standards, and Sarbanes-Oxley, require significant IT investment. Thus, financial services companies worldwide are investing hundreds of millions in revamping IT systems and financial reporting procedures in an attempt to comply with these regulations and other international financial reporting standards. These efforts comprise their attempt to conform to these new regulations and also to make their companies more financially transparent and improve their ability to assess risk. These compliance projects are generally run by managers in the finance and audit departments. However, compliance projects, particularly in the financial services industry, are exposed to failure because finance and internal audit departments do not always involve the IT director and staff at early project stages.

When IT leaders have been excluded, compliance is harder to achieve and, as a result, proposed systems may duplicate existing technology or compliance quality might be compromised. Immense time can be wasted unless compliance and IT staff work in partnership. Although finance and IT departments have been moving apart for some time, they need to work more closely together because risk specialists often have limited IT knowledge. There are lots of different pieces to the regulatory jigsaw puzzle, and so it is vital that IT departments work together with finance and compliance departments.

In a number of regions across the globe, there are numerous authorities controlling the financial services industry. These authorities insist that organisations show procedures for data retention and business continuity as well as ensure that their operational risks are being properly managed. Requirements can include provisions in relation to risk assessment and management. Risk assessment and management procedures meeting the appropriate requirements need to provide for secure data retention and recovery solutions for key IT systems.

Risk assessment and management requirements are based on a two-pronged test:

1. The probability of an event occurring.
2. The impact on core services were the event to occur.

In the enhanced risk environment that exists globally, catastrophic attacks cannot easily be discounted as comprising 'marginal' probability. Organisations must therefore seek to minimise their vulnerability and, consequently, the impact of any particular threat up to, and potentially including, catastrophic attack. Vulnerability cannot be effectively minimised if key IT systems and data do not have business continuity strategies, technologies and processes aimed at recoverability following business continuity incidents.

There are also data security requirements which state that organisations should establish and maintain appropriate systems and controls for the management of information security risks. These requirements may include:

- **Integrity** – safeguarding the accuracy and completeness of the information and its processing.
- **Availability** – ensuring that an authorised person or system has access to the information when required.
- **Authentication** – ensuring that the identity of the person or system processing the information is verified.
- **Confidentiality** – ensuring that information is accessible only to an authorised person or system.
- **Accountability** – ensuring that the person or system that processed the information cannot deny their action.

Organisations must consider the adequacy of systems and controls used to protect the processing and security of information and may wish to take account of established security standards such as ISO 17799 (Information Security Management).

Business continuity, which inevitably includes data back-up and recovery, data security, data management, business integrity and business risk and control, is essential to most financial services organisations. Organisations should also consider geographic location when considering the business operating environment of each country and whether issues such as political disruption or cultural differences may impact the provision of services and laws restricting the transfer of data across borders. In addition, organisations should consider whether local legal and regulatory requirements restrict the ability to meet regulatory obligations' such as confidentiality of customer information, access to information, and internal and external audit in a specific region. Failure to comply with any governing body in the financial services sector can, in the worse case, result in the loss of authorisation to conduct business.

3.2.1 Crime in the finance sector

The financial services sector is particularly exposed to crime including the money laundering of the proceeds of crime and terrorism. Money laundering stems from the nature of financial services – cash transactions occur on a day-to-day (if not a minute-to-minute) basis. In addition, the transfer of electronic funds increases exposure to laundering as the funds are easily and quickly moved between various accounts and organisations. Also, investment firms are at risk because liquid investment products enable illegitimate funds to be mixed with lawful monies.

All financial services organisations have an obligation to investigate and report any suspected money laundering activities. In fact, in some countries, it is a criminal offense not to maintain the necessary procedures for identifying money laundering. This can result in imprisonment if all reasonable steps have not been taken and exercised to avoid an offence.

In order to comply with applicable legislation, financial institutions must have internal controls, policies and procedures to deter criminals from choosing their particular institution for money laundering. The most effective way of achieving this is by maintaining data identification, record keeping, internal reporting and training procedures. Organisations must show at minimum

- evidence of customer identity;
- all transactions carried out by a customer in the course of relevant financial business.

These records must be retained for specific periods and organisations must provide an audit trail with easily retrievable evidence useful in money laundering investigations. Identity records and transaction records must be kept for specific terms following the end of a business relationship.

The laws relating to the financing of terrorism are well documented. It is a criminal offense for a person or company to become involved in the provision of money or other property to a person with the knowledge or reasonable cause to suspect that such property may be used for the purposes of terrorism. Suffice to say, the penalty for such an offense is a prison sentence. It is also an offense not to report a belief, or suspicion, that an organisation or individual has committed the crime of supplying financial aid to for terrorist activities.

3.2.1.1 Basel II Accord

Basel II requires financial services organisations to identify three main processes relating to credit and operational risk:

1. gathering data for reporting and analysis of credit and market risk;
2. developing data capture and reporting systems for operational risk;
3. implementing automated and standardised business processes to mitigate the risks highlighted by the reporting.

These obligations force banks to gather data they have not necessarily historically collected. This data collection usually involves accessing a host of disparate systems. Collected data then needs to be grouped and the results analysed and published. Compliance is mandatory. The lower the Basel II rating, the less competitive advantage the organisation enjoys. It therefore makes sense for financial institutions to adopt advanced approaches for credit and operational risk in order to become compliant. This also ensures that they are likely to gain better competitive advantage through greater capital release and even ratings advantages over less efficient organisations.

On the other hand, banks adopting less advanced approaches may lose competitive advantage and be less well perceived by market analysts. Organisations face huge challenges including:

- The difficulty of gathering and utilising the vast amount of information required by Basel II.
- The time consuming processes required to retrieve data stored in different organisational silos and formats.
- The gathering of information is complicated due to the sheer size of financial services organisations.
- The scale of gathered data results in a need to expand storage facilities as well as the increased use of disk for retrieval purposes.
- The greater data categorisation required to manage the increased amount of data to be made available.
- The replication of reporting systems across national and international geographic regions as well as centralised reporting systems.

- The huge IT challenge for financial institutions required to re-examine workflow technologies within the transaction processing environment.

It is difficult to overemphasise the importance of effective data processing and data retention procedures within the financial services segment. Financial data is likely to be especially sensitive. Therefore, the loss of reputation and the risk associated with breaches of data security is potentially substantial. However, the legal risks are no less serious, ranging from criminal liabilities for financial institutions and their directors for failure to comply with money laundering legislation through potential loss of authorisation to conduct investment business as a result of noncompliance with industry governance rules and requirements. Hence, these organisations must

1. Ensure the security and confidentiality of customer records and information.
2. Protect against any anticipated threats or hazards to the security or integrity of such records.
3. Protect against unauthorised access to or use of such records or information which could result in substantial harm or inconvenience to any customer.

Although financial institutions are likely to have some procedures in place already for dealing with money laundering issues and requirements, Basel II presents a need for the collation of whole new data categories and, consequently, the introduction of entirely new IT systems and processes to handle that data effectively. It follows that the loss of competitive and potential ratings advantages which may result from adopting less advanced approaches arguably makes Basel II, and the changes necessary to comply with it, the greatest challenge facing the financial services industry today.

3.3 Telecommunications companies

The sheer volume of personal data collected by telecommunications companies (telcos) means that effective and secure systems for the storage, retention and recovery of personal data assume an even greater importance than with the majority of businesses.

Maintaining data security becomes even more difficult due to the volume of data held as does

- seeking to ensure that that data is kept up to date;
- recovering all the data held on an individual;
- ensuring that no data is retained for longer than is necessary.

In addition, telcos tend to be international organisations and are therefore likely to have data protection concerns relating to data transfers outside their local country. Criminal investigation legislation examines communications interception by police and intelligence bureau. This is a response to technological advances in communications and the ever-increasing use of email and the Internet, all of which have led to criminal usage. These laws govern the power of enforcement agencies to intercept electronic data.

Effective data processing and data retention procedures are essential to telcos and ISPs (Internet Service Providers). All data retention related legislation that is applicable in other sectors of the marketplace also exists in the telecommunications sector. The main difference is that even more data is generated that is critical to both the business and legislation than with many other industries. Data storage, management and recovery present even greater practical problems and legal and regulatory compliance becomes even more onerous. The Regulation of Investigatory Powers Act, Data Protection Acts and the Privacy in Electronic Communications Regulations (EU Directive on Privacy and Electronic Communications) leave telcos in a position of having to hold vast quantities of personal data about their network and service users, and to hold more items for each call.

Data protection legislation provides a regulatory framework for the processing of personal data and is of vital importance to all businesses that process personal data. 'Personal data' is data relating to a living individual who can be identified from the data including other information which is in the possession of, or likely to come into the possession of, the business. 'Processing' is extremely widely-defined, covering almost any activity which could conceivably be associated with data including:

- obtaining, recording or holding the information or data;
- carrying out any operation on the data;
- using or disclosing the data;
- erasing or destroying data.

Businesses must observe data protection principles in their every day operations. These provide that personal data must be

- processed fairly and lawfully;
- only processed for one or more specified purposes and not excessively used for these purposes;
- not retained for longer than necessary and processed in accordance with the rights of data subjects;
- adequately protected from loss, destruction or damage;
- not transferred to countries which do not have 'adequate' measures for the protection of data in place.

Compliance with data protection legislation can be achieved by

- introducing and maintaining a comprehensive data protection policy encouraging good practice and enforcing data protection principles;
- considering whether a person's consent is required for the processing of data;
- ensuring adequate security measures are in place for the protection of data stored;
- introducing procedures to deal with requests for data;
- notifying the Information Commissioner where required.

Due to the nature of their services, telcos inevitably collect huge volumes of personal data and process that data for various different purposes including billing, marketing and the provision of services and the compiling of personal data directories.

Of particular relevance to telcos are regulations dealing with criminal investigation, data protection and electronic data management. Telecom companies naturally hold vast quantities of personal user data and hold more items every time a call is made.

The Regulation of Investigatory Powers Act (RIPA) replaces prior legislation regarding the interception of communications. RIPA is a response to technological advances in communications and the ever-increasing use of email and the Internet, which have led to misuse of communication interception techniques by both public and private organisations. It therefore governs the power of enforcement authorities and businesses to intercept electronic data.

RIPA also legislates the processing of personal data and the protection of privacy in the telecommunications sector. RIPA allows authorities to access 'communications data' held by organisations. This includes any information relating to a communication. For example, for a telephone call it would include

- identity of caller and recipient;
- their locations;
- the time and duration of the call, but not what was said during the call.

For an email, RIPA allows authorities to access data about the sender and recipient, as well as account details, locations, ISPs, IP addresses, message routing, but not the message content. With the widening of the number of bodies able to require the disclosure of communications data, there will be more instances of such demands being made on telcos.

The EU Privacy Directive introduces new information and consent needs relating to entries in publicly available directories (e.g. telephone directories) and prohibits subscribers from being charged for the exercise of the right not to appear in public directories. It also includes a requirement that subscribers are informed of all the possible usage relating to publicly available directories. The Privacy Directive has also attempted to regulate new technologies such as mobile PDA-phones and 3G technology. One example is the regulation of 'location data services'. This involves the ability of 3G phones to transmit data about their physical location to a service operator via the network. The potential benefits are obvious; however, the potential invasion of privacy is also somewhat concerning.

The extent of telcos' responsibilities in respect of data processing and data retention has exponentially increased as a result of the level of data collected from callers and other users of their services. The loss of reputation which may result from data mis-management should not be overlooked in this fiercely competitive sector. Conversely, efficient data management and recovery systems can give competitive advantage and ensure that telcos are able to extract maximum value from their vast data collections, while meeting their compliance responsibilities.

3.4 Utilities companies

Utilities companies, particularly in the competitive power and energy sectors, have additional statutory obligations with respect to data retention in the various codes and agreements under which they are required to operate as a condition of their licenses. However, there are no extra specific statutory obligations, merely a raft of agreements, bi-lateral commercial agreements and industry codes under which they operate. The legislation governing utilities mainly refers to provisions enabling the relevant Secretary of State to issue directions in times of emergency. The 'emergency' provisions generally require the utilities companies to take steps to ensure they continue to provide certain services and/or carry out disaster recovery functions, or provide fundamental 'living' necessities.

With numerous industries in this sector, these operating codes are a mandatory condition of their licenses and contain the provision of data retention and recovery procedures. Certain supply companies have to demonstrate compliance that they have the capability to operate in the relevant industry by undergoing various audit procedures and processes. One of the core competencies they need to demonstrate is the capability to transfer data to, and receive data from, other market participants, securely and effectively. Organisations need to keep the ability to retain and transfer data between the array of market participants. So, for example, a customer must be able to switch its supplier of gas or electricity and subsequently be billed correctly by its new supplier and not by its previous supplier. Such information may include technical data about the customer's meter or consumption as well as data that provide other identifying features about the customer as an individual.

Under these documents, market participants are required to transfer data in certain specified formats to other market participants including supply companies, network/infrastructure owners, meter operators, data collectors, data aggregators and settlements bodies. Some of these market participants are regulated companies and may therefore need to provide information extracted from data to regulatory bodies. Others are not regulated, but nevertheless fall within the remits of some of the industry codes. These can include network operators (i.e. companies that own/operate the relevant infrastructure, say, for transporting gas or electricity) that also have licence obligations to develop and maintain a safe and efficient system. In order to comply with these requirements, such operators need to possess secure data storage systems and contingency arrangements which would become operational in disaster situations.

Competition in the power and energy sector is based on the effective transfer of data when customers change their supplier. Compliance is now a mandatory condition of their licenses and contains provisions relating to data retention and recovery procedures. The ability to provide effective systems for the retention, recovery, and transfer of data is consequently of critical importance to utilities companies and compliance with the various code provisions on data is a mandatory condition of their licences. Customers switching suppliers would not be likely to return to a company that failed to recover and transfer their data effectively and, even leaving aside potential breach of license conditions, the associated bad publicity could be damaging in a sector which is relatively new to competition.

3.5 Public authorities and government

The public sectors' responsibilities surrounding data protection and data processing are the same as those of the private sector. However, public authorities are in a constant state of technical 'catch-up', and now, with the onset of the era of 'joined-up government' and freedom of information, they face a huge challenge. The public sector needs to implement effective electronic data management systems in order to achieve ambitious government targets and meet their statutory responsibilities.

All around the world, governments are looking at implementing Freedom of Information legislation. Freedom of Information (FOI) aims to make the information that a public body possesses quickly, easily and routinely available to the public. Basically, public sector organisations have to keep information securely available for transparent constituent access. To facilitate this, public bodies are required to devise, publish, maintain and review an individual 'publication scheme'. This is a comprehensive collection of all types of information within the public body that can be disclosed.

In order to achieve this, public sector organisations have to ensure accuracy and consistency of administrative acts, thereby reducing the risk of litigation. How can they make compliance with Freedom of Information legislation easier? Government executives do not need to figure out the business processes required to comply with the legislation (e.g. how to manage exceptions to FOI) and then develop an ad hoc IT solution. They need a solution that solves the entire problem by automating management of archiving and retrieval process and securing access to information from multiple channels for constituents.

European, Middle Eastern and African (EMEA) governments are all planning to produce 'joined up government' in the next few years. All government services are to be accessible electronically if capable of allowing electronic access. This will make improving data retention systems a necessity. Since e-government is based on the effective and efficient sharing of data, clear, appropriate and authoritative standards and the adoption of effective methods of data storage are required. Public bodies need to clarify which information standards they should adopt and have greater awareness of key products and projects that can assist and be exploited.

The stringent targets for delivery of services and electronic records management emphasises the need to act now in rationalising data management systems. Accountability means that electronic records of services are retained within effective systems and no public authority will want to suffer enforcement action consequent to the failure to comply with data protection, privacy or freedom of information legislation. For Public Authorities, and their suppliers, failure to meet these targets inevitably means loss of contracts.

Public Sector authorities need to free up precious resources for front-end services by optimising IT management and reducing costs. This means that public sector organisations need to find an IT solution that provides an integrated approach to system management to improve day-to-day operations, that ensures the continuous availability of services and minimises internal or external disruptions, as well as automating policy procedures to reduce internal IT resource requirements. By providing reliable, secure and available information sharing systems in collaborative environments, public sector organisations can

- boost information sharing initiatives where collaboration is critical: criminal justice, homeland security, health and social care;

- keep critical information continuously available for emergency response in public safety and healthcare;
- make access secure beyond firewalls for a mobile workforce (e.g. police patrols, ambulances);
- comply with legislation within national boundaries and across borders.

In government, any electronic document potentially constitutes a record accessible by the public either under existing legislation or freedom of information legislation. Since freedom of information legislation can only be as effective as the quality of the records to which it provides access, public authorities face an overwhelming need to get their data management systems and procedures in order to meet the demands of the legislation. The stringent governmental targets on the delivery of services and management of records electronically only serves to emphasise the need to act now in rationalising data management systems. Accountability demands that electronic records of services delivered electronically are retained within effective systems and no public authority will want to suffer enforcement action as a result of failure to comply with data protection, privacy or freedom of information legislation.

3.6 Managing data for compliance is just a specialised form of data management

So, the data management dilemma that all organisations around the world face today remains these regulatory requirements that cause data growth and longer retention periods. New file types, unstructured data (email, memoranda, IM) must now be treated as mission critical information. Global business forces data availability and access 24×7 across multiple time zones, planned (let alone unplanned) outages are increasingly not acceptable, and data grows, but storage remains underutilised, with static or reducing budgets. How are IT organisations to manage inconsistent processes that won't scale with infrastructure data management capabilities featuring core data management systems, including backup/restore, replication, recovery, migration, and archiving technologies, that aren't integrated.

Regulations today require companies to keep more, specific data for longer periods and selectively retrieve it on request. This increases data growth and the data dilemma organisations face today. New file types and unstructured data may need to be traceable, retrievable, and in their original state for multiple years. Hence, data remains a corporate asset that must be treated as mission critical information. Regulations require that organisations retain unstructured data and new file types including instant messages, voice mail and handheld PDA data. Capturing regulated data can be tricky, especially for previously unmanaged file types such as instant messages.

These issues focus the fact that 24×7 global companies demand rapid backup and recovery, and they have significant implications for backup/restore times and increased media consumption rates. Non-integrated backup, replication, recovery, migration and archiving products all waste resources and can result in creating multiple, redundant copies of data as well as redundant backup efforts. These failings can also increase significant media usage, cost, and management overheads, creating limited or no backup windows and high demands on recovery time objectives. The effect is that organisations are faced with uncontrollable data management costs, inefficient storage utilisation, unorganised, undiscoverable, or hidden data and large costs associated with the discovery of archived data.

The drivers for DLM or ILM centre on data growth. Although storage is relatively inexpensive, there are significant management costs beyond hardware acquisition. Management, power, floor space and so on all add up very quickly. This problem is compounded by a lack of organisation of the data and leads to very inefficient utilisation of storage capacity. Not only is there more data to manage, but also the criticality of almost all data continues to go up which increases data protection and availability needs.

Organisations can covert this situation to competitive advantage by implementing a single integrated, end to end, compliance solution. Organisations can surpass their competition by simultaneously addressing their data dilemma while not wasting time and money because they have no solution; or because they have invested in short term point solutions and subsequently are in a reactionary mode, vulnerable to unexpected, unplanned, audits.

3.7   Just plain junk data!

By extending the capability for organisations to view and understand what data they are actually storing (the ability to understand what their file based data consists of) under the control of a central facility, IT can report on such attributes as file type and file size across the entire IT domain. Then, IT can help assign a value to different data types to assist archival retention requirements, as well as control the type of file and data that users store to avoid such illegal pitfalls as copyright infringements. The other benefit of storage resource management tools is the ability to simply eliminate plain junk data, helping reclaim wasted disk space, reduce storage growth rates and backup cycles, as well as slash storage administration time, reduce risk of violating piracy and pornography laws and reduce the risk of network viruses.

Alternatively, unmanaged storage inevitably unfortunately means that storage resources become overtaxed and backup windows unnecessarily lengthened – all because users store inappropriate data on their Windows© servers. Time after time, the same things seem to appear on servers:

- end user Microsoft® Outlook® PST files;
- computer games;
- copyrighted MP3 files;
- pornography;
- graphics (wedding photos), movies and jokes;
- duplicate, triplicate and quadruplicate files;
- outdated and orphaned files;
- unlicensed software applications.

This burdens storage resources with duplicate files, outdated data, files from ex-employees or simply very old files, all sorts of offensive graphics and other media files, games, MP3s – you name it. Call it what you like, it's still 'Junk' with respect to the company mission statement.

By using a storage resource management (SRM) tool you can potentially reduce storage by 30 % instantly and then reduce data growth by a similar level. More importantly, you can ensure compliance with copyright theft laws by preventing a similar backup of potentially illegal data. SRM therefore helps reduce total costs associated with all storage expenses. The challenge is to educate organisations to see the benefit of sound storage management as against simply buying additional storage capacity every time they run out. But above all, SRM can prevent the company from being liable – which is often the case.

3.8 The bottom line – what is mandated?

The key aspects are Discovery, Legibility, Audit Ability and Authenticity. Organisations need the following across servers, end users, databases and backup data:

- a robust record retention and retrieval policy and system;
- a safe place to store data – for as long as required;
- timely discovery and retrieval of electronic records;
- guaranteed data integrity;
- an auditable process;
- to be able to demonstrate internal controls and process;
- to be able to index and journal of all archive activity;
- to be able to report in real time;
- information always available for review;
- ability to produce reports that reflect origin of data and activity.

The bottom line is that all organisations require a record retention and retrieval strategy – a safe place to store data for as long as required, which is resilient over time and compatible with legacy and future media formats and technologies. Data must be stored for fixed time periods and, in some cases, on a storage medium with specific properties such as WORM (Write Once Read Many).

During audits, organisations also require the ability to discover and retrieve electronic records in a timely manner. Therefore, efficient access to information and consistent availability is also necessary. To be effective, organisations need to be able to produce requested data, often within as little as 48 hours, or risk a more in-depth audit.

Organisations also need to be able to guarantee data integrity to protect against alteration and be able to verify originality. In other words, organisations need to ensure that the data is original and has not been altered in any way which, of course, includes newer application versions. Organisations need to be able to store original content, unalterable media, as well as new 'unstructured' file types including:

- memoranda;
- email;
- instant Messages (IM);
- other forms of digital information.

As many organisations move processes from paper-based operations, compliance regulations require their companies to demonstrate internal controls and processes in order to document what they do and how they do it, as well as demonstrate regulation adherence. Therefore, in the event of an audit, they can show who had access to the data, when, and what actions were performed – e.g. *who knew what and when they knew it.*

A system failure or lack of system visibility is not a valid noncompliance excuse. Therefore, information must always be available for review by an auditor with efficient accessibility of information and consistent availability. This also requires the ability to produce reports that reflect the origin of data and activity in real time. So, the need to manage data, the classification of data and where and when it is stored, as well as for how long, means that, amongst other things, organisations will have to manage and store data more effectively.

3.8.1 Record retention and retrieval

Regulations can have a very negative impact on IT budgets, not just from a data discovery point of view, but also simply because organisations need to store more data for longer periods of time. By failing to classify data correctly, organisations can end up storing data on inappropriate storage. After all, organisations don't want to keep their data on the most expensive storage medium for 15 years or so. In order to manage data at various lifecycle points and achieve this more cost effectively, it is important to store data in the most appropriate place. As the business value of the data decreases it migrates from more expensive faster disks, to lower cost storage, such as disks or tape.

3.8.1.1 Key technology requirements

The key technology requirements for record retention and retrieval are as follows:

• backup and recovery of all critical data types;
• tracking and support for disk, tape and optical media including unalterable types;
• archiving of new file types including email;
• automated policy-driven retention and deletion;
• high-speed content search and indexing;
• affinity grouping for line of business data;
• consistent management of storage practices and policies;
• recovery of historical application versions;
• self-service recovery for end users.

> . . . the rules of the Board may require the production of audit work papers and any other document or information in the possession of a registered public accounting firm or any associated person thereof . . . and may inspect the books and records of such firm or associated person to verify the accuracy of any documents or information supplied.
>
> Sarbanes-Oxley, Section 105(b)

There many regulations that now effect companies – and with serious consequences. The starting point for any company is to ensure that their data is backed up – protected. This is not just data on the servers, but also, and equally importantly, the data held on users' desktops and laptops which, according to IDC, equates to 60 % of corporate data. Having effectively protected the data, organisations then need to understand the differing requirements from a long term archival and retrieval point of view.

IT needs a simple and centralised way to manage the differing business requirements across their organisation for protecting data in order to achieve compliance. An important part of this is the ability to set, measure and report against clearly defined data protection service levels and equally importantly data recovery across the *entire* environment, to ensure the service levels can be met in a timely and efficient manner.

The only reasonable way to manage the retention and dele-tion of large volumes of data in complex environments is through automation. DLM and ILM strategies ensure data is stored on the appropriate medium for the correct retention period, and deleted once the period is over. A robust content search and index technology is critical to any archiving solution because it quickly enables IT to find specific data in a vast storage haystack and respond to regulatory and litigation requests for information quickly. Legal discovery costs remain a constant, so it is not just the large court cases that can heavily tax IT but any requirement, even simple retrieval requests.

By grouping like data types, IT is able to improve data tracking and streamline the retrieval process, allowing the storing of related files together on one piece of media. Automated, centrally, managed policies deliver consistent operations, reducing the risk of noncompliance and end user error. Configuration wizards, centralised administration, automated policies and intuitive schedulers, reduce human error and improve consistent operations. The more IT can enable end users to handle their own data restoration and retrieval needs, the less pressure is applied to IT department for simple deletion errors. Should they delete an online copy of data, an archived copy can be accessed through standard interfaces such as Outlook or Explorer without the need for special applications or training. So, users can access and use files, but cannot overwrite or delete the original version.

Automatically moving data from one storage location to another without changing the access path to the data allows IT to route specific data to selected media and eventually to tape for long term storage. Such policies can enable compliant archival of all electronic content. Data grouping allows related data to be relocated to the same tape or optical media for rapid retrieval. By assigning data 'retention periods', IT can dictate when data is removed, reducing the risk and costs associated with legal discovery of data you didn't need, reclaiming storage space in the process.

Becoming complaint and implementing a DLM or ILM strategy in order to create policies for record retention for long term archival and retrieval of data addresses compliance going forward. However, the ability to manage this process for historic data is equally important. This requires the ability to index existing backup data, ensuring the ability to achieve retrospective as well as future compliance.

3.8.2   Auditable process

3.8.2.1   Key technology requirements

The key technology requirements for an auditable process are as follows:

- document and publish policies;
- automated logging and journaling of events and activities;
- role-based administration;
- independent oversight of policies and practices;
- user access tracking;
- support for existing access and security;
- alignment of resources with business goals and priorities;
- organised IT resources by logical groupings ('business views');
- create and publish 'services';
- visibility into how resources are consumed;
- request of standard 'services'.

> *At all times, a member, broker, or dealer must be able to have the results of*
> *such audit system available for examination by the staffs of the Commission*
> *and the self-regulatory organisations of which the broker or dealer is a member.*
>   SEC Rule 17a-4

How does an organization create an auditable process? With document and policy templates, it is relatively easy to document and publish operation logs for internal review and external audit. However, retrieving data in its original form is useless if the application must modify it to open it. IT therefore not only needs to be able to retrieve the data, but also to access it using the appropriate application version. IT also needs to consider different access levels and control for different user classes.

Administrators, auditors, compliance officers, end users, etc. can each have their own customised access privileges appropriate to their organisational role. Different access levels allow the creation of a system of checks and balances so no single person has control over the entire system. Separate roles can be created for the storage administrator and the compliance officer. One has administrative rights for the policies used to manage data, and the other has control over the storage infrastructure.

With the need for auditable processes, IT needs to provide visibility into how resources are aligned with business strategies and how resources are consumed, as well as the cost of resource usage. IT needs the ability to discover assets via agent technology and provide a report the administrator can use to map IT assets to business views. Specific services must be definable and published for users to select. For example, due to compliance issues, a backup policy applying to specific information or applications may require data retention for seven years, whereas another may only require data retention for two years. The workflow for the provisioning of these policies can be defined and published. Based on the compliance requirements for a specific application, the appropriate service is required and provisioned.

Using DLM and ILM techniques, environments can be audited to ensure standards are met. Detailed activity, access and event logs provide a complete journal of all DLM/ILM operations and deliver a complete biography of data from the cradle to the grave. Backup and archive policies must be centrally managed – this facilitates the deployment of a compliance practice, ensures consistency of operations and makes it simple to demonstrate adherence to published policies.

3.8.3 Reporting in real time

In order to keep the compliance archive store size manageable, companies must be able to determine which data is regulated and which is not. Using sorting and prioritisation of permanent/temporary data techniques allows accurate data and available storage resource profiling. This allows the segmenting of the data within the storage environment and ultimately removing unneeded temporary data. IT must also provide consolidated, comprehensible reports on multiple aspects of the data management scheme. A single dashboard interface for compliance and backup reporting is essential for control and manageability. Using uncoordinated point products collectively makes intelligible reporting difficult, at best.

3.8.3.1 Key technology requirements

The key technology requirements for reporting in real time are as follows:

- custom and pre-defined reports;
- policy reporting;

- event and activity reporting;
- file-based reporting;
- time-stamp reporting;
- on-demand usage reports;
- web-based queries;
- sorting and prioritisation of both permanent and temporary data;
- single dashboard interface for compliance and backup reporting.

Get ready for real-time disclosure. AMR states that the current interpretation of the Sarbanes-Oxley wording of 'timely and accurate disclosure of material events' is that companies will be expected to 'disclose events that affect the business within 48 hours'.

Tech Republic, 2003

There are a number of reports necessary for IT compliance:

- **Policy reporting** – reports can be regularly created and disseminated for management review.

- **Event and activity reporting** – All events and activities are tracked, logged and journaled, creating detailed system and data biographies.

- **File-based reporting** – Granular, file information can be reported through the core infrastructure products in the VERITAS Integrated Compliance Solutions set.

- **Time-stamp reporting** – Reporting characteristics, such as the time stamp, are captured when all reports are created. This enables the ability to search back through reports based on time.

- **On-demand usage reports** – Companies must be able to produce these reports on-demand. If an auditor appears or a subpoena arrives, time is of the essence.

- **Web-based queries** – Legal and regulatory data requests can happen anytime and anywhere. Companies must be able to respond quickly, no matter where the CEO, compliance officer or legal counsel may be at the time of the request. VERITAS CommandCentral provides secure web access to necessary reports and the ability to monitor operations.

One of the best ways to be prepared for an audit is to not wait until one happens. Regular Backup does the who, what, when and where information, including:

- real-time status of backups;
- client backups;
- all log entries;
- media lists;
- media contents;
- images on media;
- media logs;
- media summary;
- media written;
- vault reports, etc.

DLM and ILM tools carry out the following:

- reporting on all data elements across the storage hierarchy;
- reporting on all data elements from 'cradle to grave';
- reporting on data lifecycle trends and detailed historical information;
- resource/service usage by logical grouping;
- defining and measuring service levels based on compliance objectives.

Keep an ongoing, detailed biography of data:

- who created it;
- who read it;
- who changed it;
- when these events took place;
- where it was created;
- where it is now;
- what policies are applied to it.

3.8.4   Integrating data management from desktop to data centre to offsite vault

Compliance solutions must leverage existing data protection resources to provide the necessary retention and retrieval technologies, while delivering additional functionality to help convert compliance initiatives into competitive advantages. Unmanaged data residing on desktops, laptops, PDA, handhelds, mobile phones, etc. must be encompassed in the total DLM/ILM strategy, thereby extending functionality to end users. SRM and reporting tools can provide insight into how the data's need for performance, availability and protection changes over time. All events and activities are tracked, logged and documented. Notification triggers can be set for specific events such as an unsuccessful operation, attempts at unauthorised access, and so on. The DLM database can also be queried to extract further information for custom reports to meet specific needs. Backup and offsite vaulting techniques provide invaluable tracking of removable media, which provides the ability to track all data, all of the time.

The business level reporting capabilities allow the information from IT assets to be reported by defined business views. The backup information is collected via an agent technology, correlated to how the company manages its business (via the views), and a portal can provide reports such as the number of restores by an application business view. Users can further filter the information to investigate, say, why a particular group is restoring the largest amount of data. In addition, specific service level metrics such as success rate or failure of backup jobs or recovery point exposure can be automatically tracked by group to ensure compliance objectives are met.

3.8.5   Challenge – the data dilemma

Never before have businesses had to face a dilemma as daunting as they face in today's fast moving critical and complex business environment. To remain competitive, organisations must rapidly scale in order to dominate their business sector, create new leading edge products and services and find new and innovative routes to market, as well as grow both organically and through acquisition by incorporating diverse systems when two or more companies merge. Organisations must also be flexible in order to ride business climate shifts, adapt to competitive threats or embrace new technologies and exploit them to remain ahead of close competition.

As companies seek new ways to surpass their competition, they start to examine their internal operations. A company's success revolves around information technology whether directly or indirectly related to business process – their revenue stream. In short, their survival entirely depends on their business data and information.

The growing complexity of information technology and IT infrastructures leads many organisations to react by installing additional server and storage capacity to counter storage and data bottleneck problems. The eventual effect is that the extra capacity is usually dedicated to specific servers and, as such, is therefore unavailable to help in other areas when needed. More staff are required to help administer the increased capacity and the infrastructure becomes increasingly disjointed and difficult to manage.

Meanwhile, business scenarios continue changing. Corporate mergers, restructuring and tactical purchase decisions are made on short notice, and proprietary IT solutions all make IT infrastructures more complex and difficult to manage. Adaptability inevitably declines and companies cannot react quickly or cost-efficiently to changing market conditions. The effect is that many companies expend too much energy reactively instead of focusing on future-oriented, proactive development that effectively deals with local and global change.

Data, information, and the means to act upon that information remains one of the most significant business assets. Instead of treating IT as a business by-product, organisations should carefully consider how to protect their IT infrastructure and how to access and share data, ensuring its availability to current and future applications. Managing information is much more than simply putting another storage device on another server and making sure it is protected from viruses.

Therefore, all organisations must plan for compliance for legacy, current and future data. In the planning, you can solve the data dilemma and in the final analysis the cost efficient approach is to solve it once, solve it right, and thereby solve it forever.

4

Data Taxonomy

Without goals, and plans to reach them, you are like a ship that has set sail with no destination.

                                                                    Fitzhugh Dodson

To understand what ILM is, and how enterprises can embark on ILM IT re-engineering efforts, it is useful to take note of the warning maxim that

If you don't know where you are going, any road will take you there ...

As we want to preclude a Sisyphean fate of perpetually developing ineffective IT storage system solutions, it is useful to ask the question: *'What should the IT storage system do*?' Here, a clear answer is that effective IT storage systems economically support the enterprise in all required present and future activities. By this, we mean an IT storage system necessarily aligns with the enterprise's objectives thereby providing a reliable storage resource that enables the enterprise to access its data as required, since a primary enterprise objective is agility to rapidly changing market conditions.

The problem with this answer is that IT organisations have historically regarded storage as a device it collectively managed. Unfortunately, the management often involved handling a lot of things at once. When enterprises required additional storage capacity, the solution was really quite simple – the IT department complied by purchasing more capacity. PowerPoint presentations that simplistically depicted storage using graphic figures unknowingly reinforced this solution by representing data as some

Data Lifecycles: Managing Data for Strategic Advantage   Roger Reid, Gareth Fraser-King and W. David Schwaderer   © 2007 VERITAS Software Corporation. All rights reserved.

sort of monoclonal entity that uniformly consumed volumetric capacity.

The rub is that enterprise storage users typically regard storage as an infinite-capacity, always-available, on-demand data repository. Clearly, a distinct possibility for misalignment between IT and the enterprise is possible and, in fact, usually occurs. To avoid this and thereby achieve alignment, IT organisations must necessarily shift their perspective of storage subsystems from one of managing *storage devices* to one of managing *data and information* in a way that is congruent to enterprise needs. The first step on this journey is for IT professionals to recognise that different data requires different access performance levels, availability, protection and migration phases, as well as different data retention periods throughout its life (as discussed in earlier chapters).

Storing new data on expensive, high-speed storage that exhibits high protection levels usually meets initial enterprise service level expectations. However, over time, stored data often becomes less critical and less frequently accessed. Such data can then migrate to a different storage resource that provides a concomitant of availability, capacity, cost, performance and protection level. This subsequent migration enables managing data retention in stages across the data's entire lifecycle. Eventually, data is no longer used. Traditionally, unused data would either be deleted to reclaim storage capacity or stored indefinitely to avoid data loss. However, regulatory requirements, legal exposure and increasing data volumes are rendering these tactics obsolete.

To better visualise the preceding discussion, consider Figure 4.1. In this figure, the various curves suggest hypothetical data criticality for different data classes – email, finance reports, etc. Not only does the criticality usually differ between any two classes at any time, it also varies over time for a given data class. Moreover, it is useful to note that data criticality does not monotonically increase or decrease for finance reports. As depicted, the criticality oscillates, probably due to periodic reporting requirements.

Figure 4.1 also depicts that before eventual deletion, data can reside on various storage types:

• online primary;
• nearline/archive;
• offline/archive.

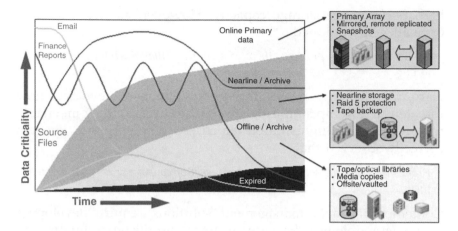

Figure 4.1 Different data has different storage needs over time

When data qualifies for migration to a new storage type depends on its criticality and its age. Finally, each enterprise has different data classes with different criticality valuations. So, lifecycle management can be regarded as an exercise in business and data analysis – coupled with technology. No single approach resolves all problems across all enterprises. There are differing approaches to accomplish the same goal. Figure 4.2 provides a view of the critical components of lifecycle management.

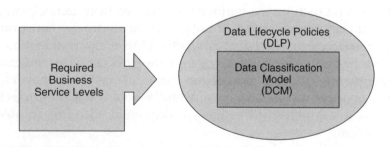

Figure 4.2 Core components of lifecycle management

4.1 A new data management consciousness level

As discussed earlier, IT organisations have historically addressed enterprise data storage growth simplistically – by providing more high performance storage capacity. However, as H.L. Mencken once allegedly quipped:

For every complex problem, there is a solution that is simple, neat, and wrong.

Here, the challenge is that the sheer volume of new data now threatens to swamp simplistic approaches because they inevitably require higher administration expenses. In this regard, Einstein correctly observed that

Problems cannot be solved at the same level of consciousness that created them.

Contemporary data management solutions require developing new methodologies because historic simplistic solutions do not economically scale. Hence, they prevent IT organisations from economically intercepting future enterprise requirements. A correct solution enables an IT organisation to incorporate heterogeneous hardware and administer its storage complex more effectively by classifying data flexibly and managing the various classifications across a heterogeneous environment with a uniform user interface providing activity automation.

Therefore, if DLM is defined as managing enterprise data from its cradle to grave across different storage media, a successful DLM solution hinges on implementing *an effective data classification strategy.*

In the next section, we remove the mystery from data classification. Data classification should not be a difficult process consuming more resources than what it is worth. However, it takes time to discover, standardise and automate the process. This assertion often generates instant antagonism from data centre managers, or operational engineers, and it is instantly noted that the project is doomed to a process of drudgery unleavened with fun and technological advances.

However, this is simply not true. The worry is the words *standards, process and automation*. Standardising the automated process does not mean that technology solutions will not be utilised. The classification creation process includes, and in most instances requires, technology. However, decisions and strategies must be made. All of these decisions and strategies involve each corporate team member and affects future architectural solution designs. The success of each one depends on the ability of the team to bring their individual experience and judgement. It is not correct to draw a dividing line along purely technical solution lines. Building an effective classification strategy requires discovering and planning.

As we stand at the threshold of a new computing and informational age, we must replace obsolete tools with more intelligence and effectiveness. We have now encountered a problem that is qualitatively different from any that we have encountered in the past; therefore, we have to introduce and use newer tools and methods as discussed in the sections that follow.

4.1.1   De-mystifying data classification

Everything factual is, in a sense, theory. The blue of the sky exhibits the basic laws of chromatics. There is no sense in looking for something behind phenomena; they are theory

Goethe

This section introduces data classification. Furthermore, it introduces the context for understanding and applying a data classification model. However, you will quickly realise that there are no right or wrong ways to classify data. Data classification is still in its infancy, and has room for additional growth. But, as you will see, data classification is a fundamental process that is the critical component to reducing overall IT costs and creating efficiencies within the infrastructure.

It is one thing to recognise a problem exists, but it's quite another to devise a solution to the problem. Over the last few years, useful ideas, terms and models for data classification have emerged that proved effective in generating useful data views. The following pages introduce those terms and ideas, and the benefits from a data classification framework, showing examples on how they can be best applied to an environment.

That said, this is the beginning of the storage strategy canvas and a process that will assist in a lifecycle management quest. Stepping through this process reveals a multitude of ways to place value on data and how to apply classification process models effectively to that data. By applying a model, true cost savings should be readily apparent by storage reduction and reallocation. More importantly, it will be shown that inefficient legacy and manual processes can be automated with a data classification plan coupled with technology; but also that automation, without a process and methodology, is a futile undertaking.

Today, it is clear that limited DLM occurs within every IT department. But, without automation, manual processing expense restricts the effectiveness of this activity. Efficient lifecycle management implementation requires automation and automation is enabled through

- classification of data, aligned with business objectives;
- federated policies governing the classes;
- technology.

Combining classified data with a set of standardised rules for handling different lifecycle scenarios enables automation. In addition, handling scenarios through automated processing drives the following:

- reduced costs from automated lifecycle management as compared to manual lifecycle management practices;
- increased scalability, solution coverage area and informational integrity;
- consistent levels of service quality.

Consistent rules, aligned with the business, enables lifecycle scenarios to unfold and data to be processed in a predictable and measurable fashion. If abstract data classification concepts cannot be reduced to identifiable objects visible to technology, the benefits of automation cannot be realised. Therefore, *business infrastructure mapping analysis* (Section 4.2.1) is a key to providing a foundation for data classification. For this reason, we have developed an enterprise data classification framework designed to empower customers to implement a lifecycle management solution effectively utilising technology. In the end, this enables enterprises to be better prepared to develop a classification strategy with realistic goals and tasks.

To assist with this process, the following items are discussed:

- defining data classification;
- classification objectives and how they apply to you;
- various approaches to data classification.

4.1.2   Defining data classification

It has been said that ILM is presently composed of 90 % process
and 10 % technology. Process, in this sense, requires categorising
and grouping data into specific categories that have similar characteristics – hence the name *data classification*. In its simplistic
form, data classification is a categorisation and grouping process
of various business rules for various data types. By tightly
coupling this process with agreed service level agreements, companies can align their business priorities with IT resources. In
the end, this ensures an organisation's most important data
assets receive the service levels they need while assessing the
business and financial impact of IT service delivery. Efficient
service delivery enables IT departments to operate more effectively by lowering the total cost of ownership and meeting
regulatory requirements. Therefore, clearly presenting a cost
and value analysis of those data services is a difficult hurdle.
However, companies need the ability to choose the lowest
cost services that meet current availability and performance
objectives.

4.1.3   Classification objectives

After defining data classification, it is important to articulate
the objectives and mission of classification clearly. Establishing a
data classification mission statement and objectives is essential in
defining the purpose of the classification model and providing
focus for the team. General classification objectives themselves
are very clear and evident, however, they can be defined further
by different organisations within a company. An example is as
follows:

1. Reduce storage-related expenses by identifying and removing
   or archiving stale or inactive data.
2. Optimise storage-related expenses through a tiered storage
   model approach.
3. Provide processes which will manage data growth more effectively through automated policies that will provide a multitude
   of retention capabilities.

4.1.4 Various approaches to data classification

We cut nature up, organise it into concepts, and ascribe significances as we do, largely because we are parties to an agreement that holds throughout our speech community and is codified in the patterns of our language. . . we cannot talk at all except by subscribing to the organisation and classification of data which the agreement decrees.

Benjamin Lee Whorf (1897–1941)

We have all seen different data classification approaches. However, experience teaches that data classification requires an orderly expressive medium. Therefore, the industry has taken an object oriented classification approach. This approach provides a defined method for representing data elements (objects) within an environment. As discussed below, this approach is both flexible and extensible. Moreover, an object oriented approach allows you a more precise data description. More importantly, this classification approach allows technology to act upon it. Just as a data object can have a state, an object orientation can define operations upon it as well because all objects have certain characteristics and behaviours.

Grady Booch, author of *Object-oriented Analysis and Design with Applications*, offers even more succinct description of an object:

An object has state, behaviour and identity.

Simplicity is in the form of the object. Furthermore, just as an object has a type it is part of a bigger picture called a class. During the classification process, we assign the data to the class because of certain characteristics. A classification class describes a set of objects that have identical characteristics.

During the classification process, we assign data objects to classes from business rules as per the *business infrastructure mapping analysis* (BIMA). The BIMA establishes and helps business unit teams develop the mission statement and objectives. The mission objectives normally define the business rules that define the classification tasks. In return, data is assigned to distinct classes for the differing data objects. As you can see, this approach is both flexible and extensible; using this approach does not limit the ways to classify data. Using object oriented approaches in this manner provides a flexible methodology to identify, characterise and manipulate the data as needed.

Now, in order to properly classify data, a method is needed. A model is a good start and assists in this process. Models are known as abstract or fictitious constructs that represent the blueprints of the classification framework. In addition, they provide a basic framework and strategy communication mechanism. Hence, planning must be deliberate.

The first principle in this process is to foresee or determine the shape of what is to come and pursue it. A data classification model is a complete abstraction of the classification process. In a sense, the model captures fundamental structural features of the listed elements. In addition, it provides and captures basic behavioural features of those elements to be acted upon. Data behaviour should have a framework that allows software technology tools to interact with data components meeting the broader goals of the business.

A data classification model facilitates the following:

- visualising and personifying data;
- capturing and communicating data analysis;
- producing methods to classify data.

4.2 Data personification

The ILM/DLM hype cycle began in its early stages with storage resource management (SRM) and morphed into active storage resource management (ASRM). Since data classification is a prerequisite to an effective lifecycle strategy endeavour, one needs the ability to identify the data assets quickly. In order to do so, technologies with SRM or ASRM capabilities should be applied.

As mentioned previously, data classification patterns capture best-practices within a specific domain. In this context, a pattern could easily solve a business problem that recurs in different contexts, such as varying data centre environments. With data classification, there are many problems such as understanding how to group data effectively into specific categories. However, grouping data effectively actually *personifies* the data, or places a face on data. As mentioned earlier, data classification assists in recognising, characterising, organising, manipulating and placing a value on data. The next few sections introduce a process to assist in this valuation.

4.2.1   Business infrastructure mapping analysis

Where and how does one start? How does one *personify* our data?
Undoubtedly, attempting to classify an entire infrastructure can
be a daunting task if done as a complete whole. Therefore, one
should begin the task by selecting a small set of critical applications. Remember, criticality is paramount due to its business
importance.

One can begin by selecting corporate email as a starting
point because of the voluminous storage it consumes. It is
also prescient to examine database applications. We live in a
database world and know that corporate data resides in a database
somewhere within the environment. Consider that Enterprise
Resource Planning (ERP) and Customer Relationship Management (CRM) relational database applications consume huge data
capacities. Data growth for these relational database systems is
already astounding and growing at nonlinear rates. Therefore, a
percentage of that data is necessarily dormant, representing an
excellent opportunity for recouping storage costs. For example,
reducing the cost of online disk storage, coupled with a reduction in offline storage, certainly provides a healthy overall cost
reduction.

So, it is clear where to start, but how does one start? Usually,
the process should begin by conducting a BIMA process. This is
the catalyst to a classification framework. The intent is to capture
the essence of the practice in a manner that is easily communi-
cated to those needing to classify data according to specific business patterns, criticality, technical reason, regulatory or compliance
reasons.

In the previous paragraphs, we noted that the BIMA quickly
defines a mission statement and data classification objectives. The
mission objectives should establish how data is classified and what
tasks should be defined. One can begin the analysis with a few
questions.

4.2.1.1   Where *can* data live (Figure 4.3)?

• What does the infrastructure look like today?
• What would it look like ideally today?

Answer: Documented Enterprise Architecture.

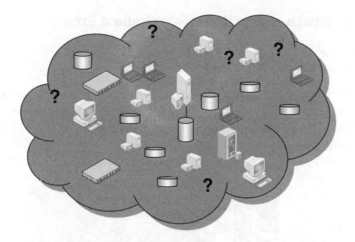

Figure 4.3   Where Can Data Live?

4.2.1.2   Where *does* Data Live (Figure 4.4)?

- What are your Active Storage Reporting capabilities?
- What are your infrastructure analysis capabilities?

Answer: Enterprise Storage Resource Management Data Profile.

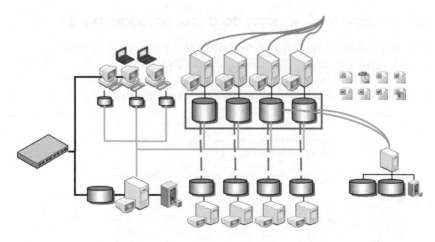

Figure 4.4   Where data can live and what data lives where

4.2.1.3 Where *should* Data Live (Figure 4.5)?

• What is the value of the data?
• What are the needs of the business?

Answer: A holistic technology solution that provides the collection, documentation and formalisation for record management guidelines, compliance/regulatory requirements and service level agreements.

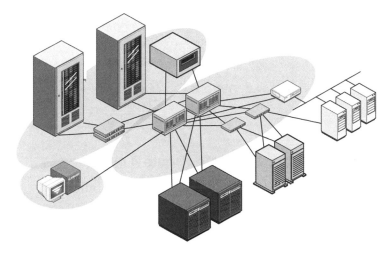

Figure 4.5 Where should Data Live?

4.2.1.4 How to view data: data classification model

• How do I manage and view heterogeneous data types?
• How do I align my data efficiently to the business?

Answer: Build a classification framework (Figure 4.6).

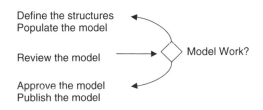

Define the structures
Populate the model

Review the model   Model Work?

Approve the model
Publish the model

Figure 4.6 Build data classification model

4.3   Classification model and framework

The above questions are typically the most frequently asked questions. Therefore, it is beneficial to see a representation or logical architectural view of the BIMA coupled with service level agreements, data classification models and lifecycle polices.

An architectural view or 'big picture' (Figure 4.7) is an abstraction of a typical model. In addition, it carefully denotes that we need to bridge the gap between IT and the business. To do that, instant visibility is needed for resources and utilisation – coupled with trending and alerting. Second, a delivery mechanism is needed for standardised processes and policy-based automation. Finally, accountability is needed with service level reporting and cost allocation.

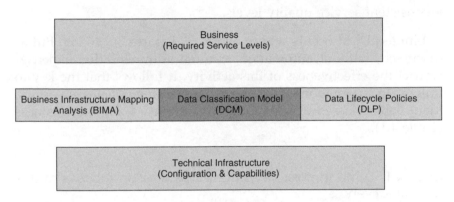

Figure 4.7 Architectural model or big picture view

Organising and representing the models in an abstract form provides an architectural and procedural focus to the overall process and methodology. However, a 'big picture' alone does not provide a clearly defined viewpoint and behavioural perspective that is easily understood. The following discussion examines the process of viewing the data and how it can presented in a manner that is both meaningful and valuable to the enterprise.

Presenting this information in a coherent and accessible form can mean the difference between every new data lifecycle project needing to relearn what is known by another data lifecycle project versus an easy methodology knowledge transference throughout the various enterprise organisations. In a nutshell, this presents a major opportunity for the DLM process.

🔒 Classification Framework Goal: *Grouping data in order to make action-able recommendations*

Combining classified data with a set of standardised rules for handling different lifecycle scenarios enables automation. In addition to these factors, using a structured approach for classifying data reduces the expenses of classifying data while simultaneously improving its value. That said, product suites (technology) assist in the automation of said rules and categorisation. Handling scenarios through automated processing drives the following;

• lower 'per scenario' management cost;
• increased scalability and solution coverage area;
• consistent service quality levels.

Limited DLM occurs within every IT department today. But as discussed, without automation, manual administration expenses restrict the effectiveness of this activity. It follows that the key to unlocking automation value is a set of federated policies governing identifiable classes of data properly aligned with the business (Table 4.1).

Table 4.1 Some factors critical to a classification model's quality and effectiveness

Aligned with the objective(s)/goals driving the classification effort
Visible to the storage and data management tools available
Extensible objects responsive to changes in business and technical conditions

Consistent rules, aligned with the business objectives, enable lifecycle scenarios to progress and data to be processed in a predictable and measurable fashion. If abstracted concepts for classifying data cannot be translated into identifiable objects visible to the technology, the automation benefits cannot be realised.

Therefore, an effective BIMA is a key element in providing a data classification foundation. For this reason this book presents an enterprise data classification framework designed to implement lifecycle management solutions effectively utilising available software products. Prior to describing a classification framework, a brief overview of the definitions is required.

Lifecycle

The period of time from data creation to final disposition.

🔒 For the purposes of building a lifecycle management, it is critical that data, information and content definitions are clear and consistent throughout the exercise. Managing lifecycle data might suggest technology is more intelligent than it is because most technology is only capable of mechanically managing *data*. The leap required to manage *information* (the meaning or intended context of the data) is fraught with risk given today's infrastructure and available technology.

Business infrastructure and mapping analysis

The process of mapping data source objects to various lines of businesses.

🔒 Linking data to lines of businesses to determine data source selection is critical in the overall methodology. Infrastructure mapping tools assist the process by identifying information brokers. Information brokers are those individuals within the company who carry links to various business infrastructure groups.

Rule(s)

A pre-defined sequence of events and actions performed on a data class for the purpose of altering the data state. Re-stated another way;

An event triggers an action to be performed on a specific data class.

🔒 *Rules* should exist only when the rule's cost/benefit analysis justifies its existence. This requires detailed understanding of the costs for implementing and supporting the rule as well as the resulting benefits. Costs typically include incremental investment in software, labour and the cost of errors within the process. Benefits will typically be measured in terms of costs reduced, expense relief through cross charge type activity and the value of reduced liability through regulatory compliance. The cost/benefit perspective should be aligned with the goals and purpose of implementing the lifecycle solution.

Data selection rules comprise criteria and actions that various data lifecycle enabled products use to select data to be copied to a data repository. These selection rules are made of criteria and actions. Data source objects can optionally be acted to be space-managed, deleted from the primary data source object (or objects) or expired in the data repository. Policies are made up of rules. Jobs use policies to archive data. There can be different categories of *data selection* rules and some are defined below.

Event(s)
Occurrences within business processes that function as a catalyst of change. Events occur when certain conditions exist signalling the existence of each event. There are two types of events: an *ad-hoc event* or a *scheduled event*.

🔒 *Events* must be identifiable and consistently trigger the same action.

Action(s)
Steps to be executed under certain conditions for the purpose of altering the state of a data class. Altering the state for a class of data is achieved through changing the *state attribute* values defined for that data class. (See definitions of attribute and below)

🔒 *Actions* must be based in the capabilities of the tools currently available for performing lifecycle management actions.

Attribute(s)
The meta-data used to define objects of a specific class or state of being within a data lifecycle model. Therefore, there are two types of attributes: *class attributes* or *state attributes*.

🔒 *Attributes* must be identifiable, meaningful and measurable for the business and technology.

State
The condition of a class of data as defined by the values certain attributes. Restated, *state* defines the existence or state of being for a class of data.

🔒 *State* must be identifiable, meaningful and measurable for the business and technology.

Data Class

A group of like data sources sharing a common set of defined attribute values. Attributes used to define are classes of data are *class attributes*.

Data classes must be extensible. This translates into the ability to adapt and extend for the purpose of responding to changes within the technical and business environment.

Data Source Object

A source of data.

Data Source Objects must be extensible and visible to the tools currently available for performing lifecycle management actions.

Figure 4.8 represents enterprise data source objects. Preceding sections described and illustrated abstract architectural views of the process. Figure 4.8 takes some example data and applies object orientation methods to assist in the classification process. Although this book is not an object orientation primer, it uses fundamental object oriented paradigm concepts. Introducing data classification in an object oriented manner assists in representing real world relationships with data. Because it represents relationships in a clear and concise manner, object orientation can be extremely beneficial because it can link and associate data using objects. Figure 4.8 is

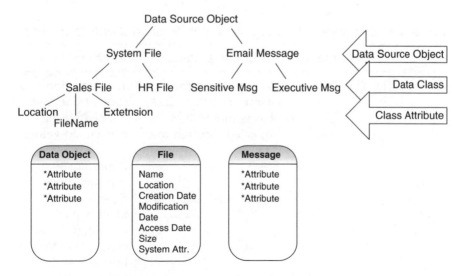

Figure 4.8 Object and Class View

therefore a representational construct of that ability and demonstrates data source object relationships.

Data classification models vary in quality and effectiveness. Table 4.1 lists some factors critical to the quality and effectiveness of a classification model. In addition to these factors, using a structured data classification approach reduces data classification expenses while simultaneously improving its value. This chapter outlines the vocabulary, conceptual points and calculated approaches for effectively classifying data.

Like all designs, a data classification model also needs to be maintained and kept current. Beyond developing the initial classification model, this discussion covers ongoing maintenance and measurement of the data classification model.

Tables 4.2 and 4.3 visually represent classes of data source objects rooted in specific states that can be applied against a given set of rules, events and actions. A data class is a classical description of a data object or a group of data objects. One then creates specific business rules for the data objects and then creates events for an action to be initiated. Data lifecycle polices can be in the form of an automated technology sub-system that provides the input into the software mechanism that moves data. The discussion in this section will assist you in further understanding this system.

Table 4.2 Example object oriented classification structure defined

Class	Data Object	Attributes/Criteria	Attribute Values
Class Name	Files/Objects	Description of the characteristics of data objects that will be impacted through this process	The condition of a class of data as defined by the predefined values. Important considerations: *State* must be identifiable, meaningful and measurable for the business and technology.

Table 4.3 Example object oriented classification structure-detailed view

Class	Data Source Object	Available Attributes	Range of Values
Class Name	Windows file	Creation Time Modification Time Last Access Time File Size File Path File Volume File Drive Archive HSM bit / reparse point Read Only System Hidden <<list all attributes>>	Date Date Date Not Modified Within <days> Modified Between <days> Created within x <days> Created between x<days> and x <days> Accessed within x <days> Not accessed with x <days>
	Exchange message	Mailbox Subject Message Recipient Message Sender Importance = Low, Normal, High Sensitivity = Normal, Personal, Private, Confidential	→ Specific Mailbox → Specific Word or Phrase → Intended Person → Importance of at most → Importance of → Importance of at least → Sensitivity of at most → Sensitivity of → Sensitivity of at least

Table 4.3 continued

Class	Data Source Object	Available Attributes	Range of Values
	File	Creation Time	
		Modification Time	
		Last Access Time	
		File Size	
		File Path	
		File Volume	
		<<Additional Attributes>>	

What Conditions Drive Data Management?
- What is a visual construct of data life cycle polices?

  - with and within service level agreements
  - incorporating BIMA

- What triggers data state changes?
- How are rules and conditions understood by the business?

   Tables 4.2 and 4.3 provide answers to those questions by representing models that can be used against data in many environments. In addition, the models can be used with a technology
solution to automate the process as well.

Example Classification Schema Based on Age   Below is a classification that meets an objective of identifying and locating data in
order to reclaim storage allocated to those particular data objects
based on age.

Data Classification Name (Class Name)
The name will represent the class of data being classified. A *data
class* is a group of like data objects sharing a common set of defined
attribute values. Attributes used to define classes of data are *class
attributes*. Or, in other words, a class is a term that describes a
group or collection of objects with common properties. In this
example we will use 'Stale' as a class of data. With that said,
we next list 'Stale' as a class of data. We would use 'Stale' as
a reference for 6 months, 6–12 months, 12–18 months and 18–24
months. The additional months can be longer than the anticipated
18–24 months, such as 24–36 months, and defined at a later time.

Class Name: Stale Six Months

Data Objects: An item of electronic data (e.g. word document, database files, etc.). Typical file types are as follows: vsd, vst, af3, txt, cvs, dwg, jpeg, pcg, tif, bmp, pcx, doc, wp, xls, ppt, vsd, tar, exe, gzip or a unique file name.

   Date Attributes: Since this class is based on a scheme of age, all the associated data objects will be based upon the *Last Created Date*, *Last Accessed Date* and *Last Modified Date*. In the case for an example company, we will use the last accessed date as the qualifier. Given the qualifier of the last access date we will provide more granular detail such as rules, methods or actions associated with the specific classes. Shown below are data attributes along with the associated values (these examples use classes based on age).

Scheme Age
Data Last Created

- Last Created Date → 6 Months
- Last Created Date → 6–12 Months
- Last Created Date → 12–18 Months
- Last Created Date → 18–24 Months

Data Last Accessed

- Last Accessed Date → 6 Months
- Last Accessed Date → 6–12 Months
- Last Accessed Date → 12–18 Months
- Last Accessed Date → 18–24 Months

In the above list for date last accessed, a rule would include all files that have not been accessed within six months and with a size greater than 100, 50, 10 and then 5 MB. Another example of the rule could be as in Table 4.4.

Data Last Modified

- Last Modified Date → 6 Months
- Last Modified Date → 6–12 Months
- Last Modified Date → 12–18 Months
- Last Modified Date → 18–24 Months

Table 4.4 Action to be taken on data objects within the class: Delete, archive, freed or storage tier placement

Class	Data Object	Attributes/Criteria	Attribute Values	Method(s)
Stale Data 6	Xyz files	Creation Date Last Accessed Last Modified Volume Drive Letter File Name File Size File Attribute	Date Values: 1. 6 Months 2. 6–12 Months 3. 12–18 Months 4. 18–24 Months Location Values: 1. Volume 2. Server: 3. <Path> Name/Extension 1. <File Name> 2. <File Ext> Size Values: 1. Larger Than 2. Smaller Than 3. Between	Delete Archive Free Tier Placement

In addition, example reports are given as well.

- Example Data Class Stale 6–12 Months
- Perspective: Location by server
- Data Class: 'Stale 6'
- Data Objects Targeted List:
- Data Objects Location/Path:

Volume Association:
Value: Accessed = 6 Months

Migration Thresholds

- Free Space Value: %
- Minimum Age: <days>
- Minimum Size: <Kbytes>
- Example Data Class Stale 12–18 Months

- Perspective: Location by server:
- Data Class: 'Stale 12'
- Data Objects Targeted List:
- Data Objects Location/Path:

Volume Association:
Value: Accessed = 12 Months

Migration Thresholds

- Free Space Value: %
- Minimum Age: <days>
- Minimum Size: <Kbytes>
- File Size: <KB || MB>

- Example Data Class Stale 18–24 Months
- Perspective: Location by Server
- Data Class: 'Stale 18'
- Data Objects List:
- Data Objects Location/Path:

Volume Association:
Value: Accessed = 18 Months

Migration Thresholds

- Free Space Value: %
- Minimum Age: <days>
- Minimum Size: <Kbytes>
- File Size: <KB || MB>

4.4 Customer reporting

This section describes the above data in reports and graphs required by most data lifecycle projects. Based on the information this chapter contains, existing reports and graphs may require customisation to accommodate these new example reporting requirements.

The reports created as a result of this exercise provide information that is used in future lifecycle management projects. Additionally, these reports serve as long-term tools for companies to use when analysing their storage environment.

The reports are organised in the following sections:

• Summary reports – These reports give aggregate environmental information pertaining to servers, users and files.

• Detailed reports – These reports allow users to get specific information on individual servers, users or files.

• Summary graphs – The graphs that can be customised to provide trending information that facilitates DLM decisions and provides better visibility into the current storage environment.

Throughout this discussion, we use the term 'old data' where 'old data' is categorised by the file access dating. The actual age of 'old data' is a configurable set of data points.

For initial report construction, 'old data' categories can comprise the following:

• last accessed between 3–6 months ago;
• last accessed between 6–12 months ago;
• last accessed between 12–24 months ago;
• last accessed >24 months ago.

4.4.1 Summary reports

Summary reports emphasise 'old data' information. To present the information a lifecycle project requires, three summary reports, described below, are required:

• Server Summary Report
• File Summary Report
• User Summary Report

For all reports, a scanning strategy enables the reports to display data on, say, the oldest 10,000 files. Consequently, these reports are not an all-inclusive view of a complete storage environment. Rather, the reports provide information useful in creating 'data classes' for lifecycle efforts.

Furthermore, reporting does not include files that are not impacted by a DLM solution. Storage not reported on is termed 'white space'. Here, white space is defined as data objects not impacted by the solution (e.g. device files, DLL files and data intentionally not collected (i.e. beyond 10 000 files)).

4.4.1.1 Example server summary report

Report Name: Server Summary Report
Report Date: dd/mm/yyyy

Server Name	Total # Files	Total Size Used(MB)	MB Old Data	%Old*	# Old Files	% Old*
Server A	99 999	99 999.00	99 999.00	99	99 999	99
Server C	99 999	99 999.00	99 999.00	99	99 999	99
Server B	99 999	99 999.00	99 999.00	99	99 999	99
Totals:			**99 999.00**		**99 999**	

Initial sort field: % Old Data
Grouped by: Server Name

4.4.1.2 Example file summary report

Report Name: File Summary Report
Report Date: dd/mm/yyyy

File Type	Total # Files	Total Size Used(MB)	Amount of Old Data(MB)*	% of Old Data
File Type A	99 999	99 999.00	99 999.00	99
File Type B	xxxxxx	xxxxxxxxx	xxxxxxxxx	99
File Type C	xxxxxx	xxxxxxxxx	xxxxxxxxx	99

Sorted by: This report can be sorted by any of the columns in the report. For example, this report can be sorted by *Total # Files* or by *% of Old Data* depending what information the user would like to highlight.
Grouped by: File Type

4.4.1.3 Example user summary reports

Report Name: User Summary Report
Report Date: dd/mm/yyyy

User Name**	Total # Files	Total Size Used (MB)	Amount of Old Data (MB)*	% of Old Data*
User A	99 999	99 999.00	99 999.00	99 999.00
User B	xxxxxx	xxxxxxxxx	xxxxxxxxx	xxxxxxxxx
User C	xxxxxx	xxxxxxxxx	xxxxxxxxx	xxxxxxxxx

Sorted by: This report can be sorted by any of the columns in the report. For example, this report can be sorted by *Total # Files* or by *% of Old Data* depending what information the user would like to highlight.
Grouped by: User Name

Notes
**In order to identify functional/application User IDs in an environment, consider the following:

- Most functional accounts are created with the three-letter prefix 'abcxxx'.
- There might be functional accounts that need identification by filtering out all IDs beginning with the appropriate letters, numbers or combination and then searching for unique User IDs with more or less than seven characters.

4.4.2 Detailed reports

4.4.2.1 Server scan detailed report

This report enables data lifecycle administrators to understand which servers are not reporting scan information. Essentially, this report ensures that appropriate data is collected from all servers.

Report Name: Server Scan Report (Servers with no scans within 7 days)
Report Date: dd/mm/yyyy

Server Name	Date Last Scanned
Server A	dd/mm/yyyy
Server B	dd/mm/yyyy
Server C	dd/mm/yyyy

Sorted by: Date Last Scanned
Grouped by: Server Name

4.4.2.2   Server detail report

The Server Detail report gives specific server information. The scanning strategy dictates this report shows the 10,000 oldest files. The user can sort records in this report based on the size of file, file type or file size.

Report Name: Server Detailed Report for Server A
Report Date: dd/mm/yyyy
Server Name: Server A

Old Data Range	Total # Files	Total Size Used (MB)	Old Data (MB)	%Old*	Old Files(#)	% Old*
3–6	99 999	99 999.00	99 999.00	99	99 999	99
6–12	99 999	99 999.00	99 999.00	99	99 999	99
12–24	99 999	99 999.00	99 999.00	99	99 999	99
24+	99 999	99 999.00	99 999.00	99	99 999	99

4.4.2.3   File detail report

The File Detail Report gives specific file information. A specific scanning strategy suggested earlier shows the 10,000 oldest files. Users can sort records in this report based on the size of file, file type or file size.

Report Name: File Detailed Report
Report Date: dd/mm/yyyy
FileType (.log), Access Date Range 3–6 months

File	Location	Owner	File Size (MB)	Last Access Date
File Name A	/path/path	userid	99 999.00	dd/mm/yyyy
File Name B	xxxxxx	xxxxxx	xxxxxxxxx	dd/mm/yyyy
File Name C	xxxxxx	xxxxxx	xxxxxxxxx	dd/mm/yyyy

Total Space Used (MB): 99 999.00

FileType (.log), Access Date Range 6–9 months

File	Location	Owner	File Size (MB)	Last Access Date
File Name A	/path/path	userid	99 999.00	dd/mm/yyyy
File Name B	xxxxxx	xxxxxx	xxxxxxxxx	dd/mm/yyyy
File Name C	xxxxxx	xxxxxx	xxxxxxxxx	dd/mm/yyyy

Total Space Used (MB): 99 999.00

. . .

FileType (.tar) Access Date Range 3–6 months

File	Location	Owner	File Size (MB)	Last Access Date
File Name A	/path/path	userid	99 999.00	dd/mm/yyyy
File Name B	xxxxxx	xxxxxx	xxxxxxxxx	dd/mm/yyyy
File Name C	xxxxxx	xxxxxx	xxxxxxxxx	dd/mm/yyyy

Sorted by:
Initial Field Sort: File Type, File Size
These reports can be sorted by any of the columns in the report.
For example, they can be sorted by *Path Name* or by *Last Access Date* depending what information the user would like to highlight.
Grouped by: FileType, Last Access Date

4.4.2.4 User detail report

The User Detail Report gives information on the data associated with specific users and how they utilise tier one, two and three storage.

Report Name: User Detail Report for User A
Report Date: dd/mm/yyyy
Access Date Range 3–6 months

File Name	File Type	Server Name	Location	File Size (MB)	Last Access Date
File A	xxxxxx	xxxxxxxxxxx	/path/path	99 999.00	dd/mm/yyyy
File B	xxxxxx	xxxxxxxxxxx	xxxxxxxxx	xxxxxxxxx	dd/mm/yyyy
File C	xxxxxx	xxxxxxxxxxx	xxxxxxxxx	xxxxxxxxx	dd/mm/yyyy

Access Date Range 6–9 months

File Name	File Type	Server Name	Location	File Size (MB)	Last Access Date
File A	xxxxxx	xxxxxxxxxxxx	/path/path	99 999.00	dd/mm/yyyy
File B	xxxxxx	xxxxxxxxxxxx	xxxxxxxxx	xxxxxxxxx	dd/mm/yyyy
File C	xxxxxx	xxxxxxxxxxxx	xxxxxxxxx	xxxxxxxxx	dd/mm/yyyy

Sorted by:
Initial Field Sort: File Size (MB)
These reports can be sorted by any of the columns in the report. For example, they can be sorted by *Files Type* or by *% of Old Data* depending what information the user would like to highlight.
Grouped by: Access Date

4.4.2.5 Waste detail report

The Waste Detail Report gives information on the data that can be eliminated from the environment (Duplicate files, business inappropriate data, etc.).

Report Name: Waste Detail Report
Report Date: dd/mm/yyyy

File Name	File Type Name	User (MB)	File Size	Last Access Date	% of Old Data*
File A	xxxxxx	xxxxxx	99 999.00	dd/mm/yyyy	99
File B	xxxxxx	xxxxxx	xxxxxxxxx	dd/mm/yyyy	99
File C	xxxxxx	xxxxxx	xxxxxxxxx	dd/mm/yyyy	99

Data Class (Duplicate Files)

File Name	File Type	Server Name	Location	User Name	File Size (MB)	Last Access Date
File A	xxxxxx	xxxxxxxxxxx	/path/path	xxxxxx	99 999.00	dd/mm/yyyy
File B	xxxxxx	xxxxxxxxxxx	/path/path	xxxxxx	99 999.00	dd/mm/yyyy
File C	xxxxxx	xxxxxxxxxxx	/path/path	xxxxxx	99 999.00	dd/mm/yyyy

Data Class (Media Files)

File Name	File Type	Server Name	Location	User Name	File Size (MB)	Last Access Date
File A	xxxxxx	xxxxxxxxxxx	/path/path	xxxxxx	99 999.00	dd/mm/yyyy
File B	xxxxxx	xxxxxxxxxxx	/path/path	xxxxxx	99 999.00	dd/mm/yyyy
File C	xxxxxx	xxxxxxxxxxx	/path/path	xxxxxx	99 999.00	dd/mm/yyyy

Sorted by:

Initial Field Sort: File Size

These reports can be sorted by any report column. For example, they can be sorted by *Files Type* or by *% of Old Data* depending what information the user would like to highlight.

Grouped by: Data Class

4.4.3 Summary graphs

4.4.3.1 Storage consumption and trending

Figure 4.9 depicts three things:

1. Capacity – Current storage capacity in the environment.
2. Baseline – Storage consumption and trend for future consumption that estimates when the current storage capacity is consumed.
3. Actual – Storage consumption and impact of a data lifecycle project (storage cleansing and/or policy-based effects). DLM should provide a new storage consumption trend that estimates when the current storage capacity is consumed.

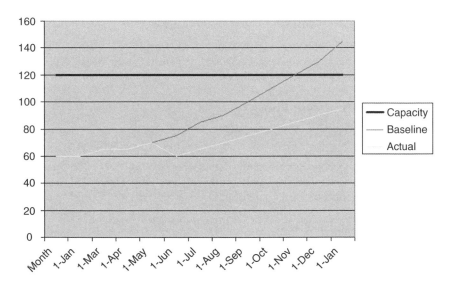

Figure 4.9 Summary graph depicting storage capacity and consumption

4.4.3.2 Data analysis 'old data'

Figure 4.10 shows the amount of existing data in an environment that falls within the *old data* definition. The figure depicts the amount and age of the data being reported. For example, the red start on the figure indicates there is ~22 600 GB of data that has not been accessed in 31 weeks. The analysis that can accompany the report can provide an estimate of how much old data is in the environment and can be impacted by an effective solution.

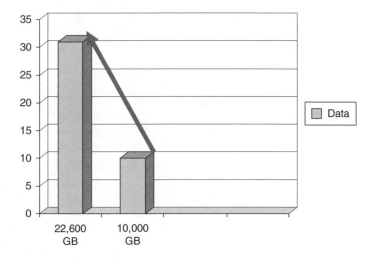

Figure 4.10 Amount of existing data defined as 'old data'

4.5 Summary

As we have discussed, data classification is the foundation to a successful data lifecycle process and methodology. By carefully aligning business goals and objectives to your current and future lifecycle projects it will assure a company's competitive advantage over their competitors. Personifying the data is a clear choice for meeting critical company objectives and missions. The example models and reports given in this chapter should provide you with enough details to make a more practical decision.

As technology advances improved mechanisms will be come available to assist in this endeavour. Inefficiencies will soon become efficient and impenetrably complex choices become more obvious. Companies that make intelligent decisions about how the data is organised, represented, located and acted upon will soon see financial benefits and competitive advantage in the new digital age.

5
Email Retention

Email has rapidly become the main communication means between individuals and organisations – whether peer-to-peer, business-to-business or business-to-consumer. But unlike other forms of communication, email has a tendency to encounter service interruptions, disruptions and outages. As IT departments simultaneously wrestle with email availability, as well as spam and viruses threatening the organisation, corporate compliance and information security managers are grappling with email leaving and entering the organisation.

It has been suggested that over half of an organisation's business critical information could be stored in the corporate email repository. Many users now view their email system as a filing system keeping everything, forever. The result is that the management, storage, archival and retrieval of this information is an onerous task.

Oddly enough, when it comes to communicating sensitive information, email is usually the preferred format. Given its inherent insecurity, this is a huge surprise. But it is the ease with which we can share information that gives email its appeal. Formally writing a company letter or an internal office memo to move than 50 people can be far too much like hard work. Making a telephone call that remains hearsay a week from now is less appealing than banging out an email. This is all despite the fact that some of the more traditional communication forms offer much more inherent security than numerous email systems. Fundamentally, email requires

Data Lifecycles: Managing Data for Strategic Advantage   Roger Reid, Gareth Fraser-King and W. David Schwaderer   © 2007 VERITAS Software Corporation. All rights reserved.

a holistic management approach that heeds technical, procedural and cultural issues.

5.1 Email management to achieve compliance

Undeniably, one of the most important enterprise applications is the corporate messaging system, which is typically email. Also undeniably, email has become one of the most critical applications. But, with compliance and storage issues, as well as application performance and availability, email is also an IT management nightmare – particularly as it relates to compliance legislation, policy and regulation.

There are three types of compliance:

1. regulatory (driven by federal, country and local agencies);

2. corporate governance (driven by internal executives and legal departments);

3. codes of practice (driven by industry vertical markets).

The first type has very strict penalties associated with noncompliance in the form of hefty fines and/or jail time. Corporate governance, however, deals more with setting up best practices to protect a company from potential legal and human-resources issues.

Corporate IT departments encounter numerous myths about what they must do to achieve compliance. Some storage vendors claim that backups constitute archiving. Other vendors approach end users and tell them that they have 'certified solutions', even though only one regulation (DOD 5015) provides any type of certification. Therefore, it can be very confusing for IT managers to determine what they must do to have the best email management practices, as well as the ability to develop methodologies that allow them to achieve both regulatory and corporate governance compliance.

When it comes to managing email for compliance, some best practices and standards have been established. Information from consulting firms is available for organisations striving to become more knowledgeable on the best practices for electronic data management. When all of this information is boiled down, however, there are four key components that companies must pay attention to in order to ensure compliance:

- **Email must be tamper proof:** Email must be password protected, read only, non-delete-able, encrypted and digitally signed, and exist in a closed system online and offline.
- **Email must follow the defined polices of the business:** Policies include what email is archived, where email is archived, how long email archives are retained and how email is protected.
- **Email must have full audit ability of access and movement:** Email must have the ability to be audited by a third party.
- **Email must be fully indexed and provide full search capability:** Specifically, email archiving must be indexed based on capturing standard RFC-822 header information.

With these components in place, corporate IT departments can also develop a scalable email infrastructure that is easy to manage. Additionally, these email best practices allow IT managers to manage more with less, saving valuable time, money and resources.

5.2   What is archiving?

With various storage vendors defining 'data protection' to mean anything they want as it pertains to their products, many customers are confused as to what, exactly, backup and archiving measures entail.

- **Backup** is the activity of copying files or databases so that they are preserved in case of equipment failure or other catastrophe. It is usually a routine operational process of large businesses with mainframes or smaller business computer administrators. Backups are usually performed once per day and capture data as it exists at the end of the day. There are no criteria with backups that require an original copy of the data be preserved, so the authenticity of the data being captured is not of primary concern. The recovered data or recovered computer will reflect the state of the data, typically from the evening at which the backup was taken. This means that as a file changes over the course of a day, only the version of the file captured at the end of that day can be recovered. Although backup is a highly recom-mended and often-required part of data management operations, it shouldn't be relied upon if archiving is what's really required by the businesses. Often, processes for archiving and backup need to function in tandem.

• **An archive** is a collection of computer files that have been packaged together for backup, to transport to some other location geographically distant from the computer. Archiving creates an exact copy of data that never changes and is preserved for any number of reasons. This 'exact' copy provides authenticity to anyone who would need to know that the version of the file is exact. Proper archiving captures a copy of a file each time it is saved such that each version is captured. Additionally, multiple copies may be made so that they can be moved to some permanent storage device to keep for long periods of time, and the original copy can be deleted to save on primary storage space.

5.2.1 Email archiving requirements

When it comes to best practices for archiving email, there are four key factors to keep in mind when choosing an email archiving solution.

• **Discovery**. Information must be easy to access and consistently available in order to successfully meet legal discovery challenges from regulatory committees. In the case where companies do not need to meet regulatory requirements, having data readily assessable allows end users to search for information which they may have lost or accidentally deleted, alleviating calls to the help desk.

• **Legibility**. Information must be able to be read today and in the future, regardless of technology. One problem with backup systems is that they need to exist for years in order to recover tapes created on old systems with old backup software. As a result, when selecting an archiving technology, companies should look for solutions that are based on open systems, in the event that their email applications should change. For example, if companies migrate from Microsoft Exchange to Lotus Notes (or vice versa), they must still be able to obtain archived email.

• **Audit-ability**. An email archiving solution must have the ability to allow third parties to review information and validate that it is authentic.

- **Authenticity**. Information must meet all security requirements, account for alteration and provide an audit trail from origin to disposition. An audit trail can track any changes made to an email; anyone performing a search is informed when a file is altered and by whom.

Whether or not IT aims to create an email archiving system for corporate governance or regulatory compliance, if an email archiving solution meets these four criteria, then any piece of information captured in the archive is authentic, ensuring companies are confident that the material they present is authentic evidence.

As IT pressures continue to require keeping more data for longer time periods due to compliance efforts combined with the sheer increase in electronic data, email management can seem like an IT professional's worst nightmare. Once IT makes a knowledgeable distinction between email backup and archiving, however and is aware of the best practices for archiving, challenges such as new compliance regulations and storage and recovery issues can be managed with success.

5.3   How should organisations manage their email records?

Viruses and spam email are well known to IT administrators, but organisations face a far greater risk that extends beyond the bits and bytes of information technology. The subject of email retention and compliance regulations has become a hot topic over the last couple of years. Until fairly recently, legal departments still advised businesses to delete all email. But once an email gets outside a company's firewall there is no knowing how many copies may be held – copies that, if deleted, the company has no way of knowing what the possible exposure may be if they came to light at a later date. (There was a marvellous case where an employee received a rather lewd email from his girlfriend and sent it to a couple of 'friends', within day or weeks the email had travelled the world!) This is a good example of how a single recipient email can go global within a few weeks because email can proliferate so easily.

In effect, this means that when you view messaging, business or otherwise, notwithstanding the regulations which refer to the retention of data, organisations can never presume anything is unimportant. This definitely applies to email, instant messages and text pages. All these can be considered as business transactions or even evidence for investigators under laws like Regulation of Investigatory Powers Act (RIPA), as well as the regulations and standards that require all forms of data to be stored intelligently and have the ability to be retrieved and proven to be credible should they need to be produced as evidence in litigation. Organisations, therefore, require practical ways of keeping email messages that will give them sufficient reliability as evidence if required.

> *A Federal District Court Judge ruled that [a financial institution] must search its archives and restore a number of email messages the company removed from its system, in response to a plaintiff request, even though the bank claimed the retrieval of the emails would cost at least $175 000.*
>
> New York Times, 17 May 2003

They failed to comply, were held in contempt of court and fined some $500 000. At least one email has never been recovered (the correct number of missing emails is unknown, but is at least one). The organisation in question failed to maintain email records as a corporate entity and gave end users too much control on what they did and did not keep. Hence, it is no longer sufficient simply to notify all employees of a compliance issue and expect that employees will then retain and produce all relevant information when required.

Organisations must also ensure that all data is discoverable and that relevant information, especially electronic information, is retained and able to be produced in a timely manner when requested. One of the primary reasons that electronic data is lost is ineffective communication with information technology personnel. In the UBS case, legal council failed to let the IT department know that there was a pending court case. So backup tapes continued to be recycled as normal. Therefore, it is important that organisations ensure that IT departments safeguard all potentially relevant backup tapes, and make certain that backup data tapes, for example, are not inadvertently recycled. This particular case also advances the argument that organisations should not give end users the power to delete email. Arguably, then, you could say that even viruses should be kept just in case a court case requires the evidence.

5.4   Email retention policies are for life – not just for Christmas

To make matters even more complex, there are a number of data protection codes and employment practices to which organisations must adhere. In conflict with these codes is terrorism and criminal legislation that empowers organisations to intercept communications in all its forms and supply information to appropriate law enforcement agencies. The problem with the interception of any type of communication is the possible violation of the basic human rights legislation. Since email is a form of data processing, employers must consider whether the interception of email is a violation of these human rights laws. In other words, does the monitoring of work email unnecessarily intrude on an employee's privacy?

There are, however, employment practices guidelines or codes that help organisations set policy to comply with the provisions of the data protection acts. Having reviewed the legislation, organisations should consider the following points:

• Should email monitoring be limited to data traffic rather than actual email contents?

• Should organisations consider spot checks rather than continuous monitoring?

• What risk areas need more monitoring than others?

• What steps can organisations adopt to reduce the extent to which extraneous information is made available to any one person?

• How can organisations consider the necessity of monitoring versus the intrusion into individuals' privacy?

These employment codes provide benchmarks that employers are expected to meet in order to comply with specific data protection acts. It is clear that, in any prosecution or other enforcement action, the policies an organisation sets will be considered depending on the organisation's ability to establish, document and communicate a policy on the use of electronic communications. In other words it is better to have a company policy on electronic communications than to ignore the issue, even if some of the policies set are contrary to some legislative acts (all companies should ensure they take independent legal advice on this issue). So it is essential that organisations have an acceptable use policy in place that is made known to all their employees.

5.5 How companies can gain competitive advantage using compliance

Organisations that efficiently and quickly address compliance – those that remove the internal barriers between IT and the lines of business and take a proactive approach towards compliance – are the ones that will retain competitive advantage and increase their business efficiency. An audit request can jumpstart organisations into examining email archiving and data management, However, reacting to an email audit request before addressing the data management issue tends to suggest being too late.

Most organisations balk at facing up to the issue as they think that there will be a requirement to throw out their existing IT infrastructure and begin again – however, this is certainly not the case. What it does mean is that organisations have to understand what type of data they have and what the requirements are affecting data, together with what content types have to be retained and for how long. They must then put in place an auditable process to store, manage, retrieve and report on all these information assets as outlined in the particular regulations within the vertical market or geographic region.

New global and market specific regulations set new requirements for companies to comply with regarding their essential business data. The laws govern the process and methods that organisations use to create, store, retain, and access records. It all comprises storing more information longer, and with greater risks for failing to do so.

Email retention is a relatively new compliance issue. Early 2004 Financial Services Authority investigations repeatedly used email as evidence, revealing the extent to which emails are used in litigation.

It is important that the data can be archived and retrieved successfully. Generally compliance issues are seen as being predominately aimed at the public sector and financial services industry. But this is not the case. Around 75 % of all data discovery requests in legal cases across all vertical markets are now for email. Litigation extensively uses email and it is likely that organisations from every vertical market sector will eventually find themselves involved in lawsuits where email will provide valuable evidence.

Although most organisations now understand the necessity to save data, including email, what is not clear is whether these organisations actually understand how complicated it can be to retrieve a specific subset of emails and the costs associated with these requirements. There are many examples, primarily in the USA, where organisations are required to retrieve emails at exorbitant costs. There are numerous stories of email retrieval costing between $100 000 and $500 000 simply to obtain the required information from a number of sample backup tapes.

The consequences of nonretrieval are even more profound. Some companies believe that it is cheaper to pay the fine levied through nonretrieval rather than pay to find and retrieve the data. This development prompts two fundamental questions:

1. If organisations are not going to retrieve email for compliance regulation, why are they backing them up and wasting money storing them in the first place?
2. How long will the regulators and the legal profession accept organisations being contemptuous of the regulations before the fines levied ensure compliance?

It would be naive to continue believing that enterprises do not have to get their email houses in order.

5.5.1 Compliance makes good business sense

To makes things easier for organisations, there are a number of really good reasons for archiving properly. However, organisations need to learn that as most legal requests for email relate to a specific set of emails, simply using backup tapes as an archive for retained emails is not an adequate solution to compliance requirements. If it is cost that worries the enterprise, the email archiving system footprint is both small and certainly pays for itself in one or two occasions where emails must be retrieved at a granular level. In addition to simplifying and reducing the cost of the discovery process, an archive eliminates the problems of full mailboxes and the battles between email system administrators, compliance officers and end users.

Archiving, of course, is not without problems. Most business issues are resolved at the business level. However, there are numerous issues seen as purely relating to the IT department. So, on occasion, email retention policy is laid at the feet of IT without the business deciding at what point an email should be archived. Should it be on arrival into the organisation before it is delivered to the recipient, or after a period of time, say 30 days, assuming the recipient has not deleted it?

From what we have discussed so far, we can assume that the former is the starting point for a sensible policy for email retention. Access to the email archive is essential to any email archiving solution. End users should only have access to their own archived emails; they should be able to manage their archive but should not be able to delete emails. This is especially important from a compliance perspective, and must form part of the email retention policy.

Compliance is not a single-event activity but an ongoing process. As the number of regulations and legislation grows, an increasing number of organisations are required to retain business emails. Employees cannot be expected to know which emails need to be retained and which can safely be deleted. Even where retention is not required at the moment, it is still preferable not to leave email management in the hands of staff. A case in point is the UBS case where end users may have attempted to obscure their tracks by deleting emails that should have been kept by the company and been subpoenaed by the courts.

Despite the obvious risks, the sensible approach is to retain business email, regardless of whether it is currently required for compliance, as it often forms proof of events that took place or electronic conversations. It can be difficult to decide what is business email and needs to be retained, and which emails can safely be deleted, such as spam.

To make matters more complicated, there are various retention periods for different pieces of legislation. Each regulation specifies a different length of time, as does the form in which it needs to be made available to the regulator. It is estimated that there are more than 10 000 differing regulations in the USA alone and probably more than 15 000 in Europe, Middle East and Africa (EMEA). The chances are, therefore, that any one organisation will be subject to a number of differing regulations, best practice requirements or laws under which email must be retained, each with different retention periods.

Some organisations choose to retain emails beyond the retention period because of the value of the information held within individual emails, but this could be risky. Not only because there are limitation laws which restrict the length of time that data can be held, but also because of the balance between the value of the data to the organisation and the risk of it being used in litigation. Therefore, organisations need to balance the value that can be gained from an email against its danger when deciding how long to retain an email beyond its retention period.

Having an email retention policy is better than not having one. By reducing instances of noncompliant emails, the regular review of emails becomes a less difficult task. There are now products that can index data and track it throughout its life, allowing IT to tell an auditor or regulator who created the document, who amended it, where it has been stored, where it is stored now and when it was last retrieved, etc. This ability allows an organisation to review and store email as appropriately. This proves to the regulator that the organisation has implemented policies to deal with email.

There are a number of email retention approaches:

1. Retain all emails, including spam, to ensure compliance. The downside here is the size of the archive and also the impact on searching the archive.
2. Filter out nonrelevant emails, by using an external spam filter with the ability to check rejected emails to ensure that they are spam.
3. Keep all emails, categorising them according to content and giving different retention periods to each category depending on limitations laws and relevance to the business.

Basically, organisations need to regard compliance as a long term issue and ensure they invest in solutions that scale as regulations change or increase in numbers. It seems likely that enterprises have not seen the last corporate scandal which will, inevitably, mean tighter laws and additional legislation.

5.6   What laws govern email retention?

As mentioned earlier, a plethora of guidelines, regulations and legislation dictate what email policy should be. There are plenty of laws (Sarbanes-Oxley and SEC/NASD) which may affect a large number, if not all, countries around the world. In fact, Basel II is a good vertical market example, where its 3rd principle directly refers to Sarbanes-Oxley.

Basically, data retention laws are pretty universal. But, of course, there will be variations to the laws and retention requirements discussed here. Companies addressing compliance issues should refer to independent legal council for further advice. Compliance regulations are liable to change and nothing in this book should be taken at face value without first referring to a legal entity.

5.6.1   How long do we have to keep email records?

This is the most frequently asked question. Interestingly, few countries currently have specific laws explicitly stating that emails must be retained. However, what the law does state is that it is necessary to ensure that certain documents are retained. Documents, which are basically anything in which information of any description is recorded, can be requested or brought as evidence in court. Nowadays, emails are considered documents. The medium for the recording of the information is largely irrelevant and the requirement for emails to be produced as evidence for both civil and criminal cases is on the increase.

There are basically four types of documents when considering document or email retention:

1. best practice standards (ISO 17799, PAS 56, etc.);
2. documents that must be retained for general legislation (International Accounting Standards etc.);
3. documents required for vertical sector-specific regulations (e.g. FSA regulations, manufacturing standards, etc.);
4. documents that may support a civil or criminal action in the future (sex discrimination, constructive dismissal, breach of contract, etc.).

For most vertical markets, general legislation and specific regulations or policies already exist for paper-based documents. There may even be updated versions for electronically-held data. These regulations reflect the appropriate legislation and define corresponding retention periods. What organisations need to ascertain is whether these document requirements are contained in emails (including attachments) and set policy accordingly. As well as general legislation, the requirement for industry standards is pretty straightforward. However, simply retaining documents that may prove useful in the defence or prosecution of civil or criminal legal proceedings is more difficult to define.

The law does not mandate organisations to retain records merely because they may prove useful in legal proceedings. However, as in the case of UBS Warburg, courts can become quite upset when an organisation fails to do so. Hence, it makes sense to do so when possible. Not only is there an impact when organisations cannot produce documentary evidence as required, but it certainly can adversely affect an organisation's ability to prosecute or defend an action. So the importance of following the local Limitation Acts, for the specific country to which the data applies is painfully clear.

The tricky bit is setting policies for a single piece of data when there may be a number of rules or regulations that require fulfilling. For example, Data Protection Acts can cause some confusion in relation to retention periods for emails. In some cases, the act may suggest some, or all, of an employee's email records should be deleted when they leave the organisation. This, of course, clashes with the principle of keeping all ex-employee emails in case the organisation needs to defend a future legal claim. In these instances, the limitation period for potential actions therefore become the logical retention period. To comply with the Data Protection Act in this case, it is prudent to ensure that an ex-employee's email box is saved with the appropriate security on the most appropriate storage – this may even be an offsite tape vaulting storage.

5.7   Write once, secure against tampering

It's all fine and dandy to ensure that end users do not accidentally (on purpose) delete email, say relating to a sexual discrimination case, but it is just as important that they are prevented from retrospectively altering evidence after the fact. This means that all email brought in evidence in a court case need to be valid, or have evidential weight. This means that organisations must be able to prove that the information presented can be considered to be an accurate and true record that has not been tampered with during its lifecycle.

There are codes of practice that describe how organisations can demonstrate to a court that the contents of an email, or document, created within a computer system has not been tampered or changed since the time of storage. The following five principle of document storage are based on the UK Code of Practice for Legal Admissibility of Information Stored Electronically (BIP 0008 2004, 3rd edition):

1. An organisation should adopt a documented *Information Life-cycle Management policy*, ensure that it is regularly reviewed every six months and that the policy responsibilities are clearly defined. The policy should cover the storage of data and information, including email, how each type of data type is stored, the appropriate security level and the correctly defined retention and deletion periods.

2. Organisations should implement an *Information Security Policy*, together with a disaster recovery, or better still, a business continuity strategy, plan to 'show duty of care'. To do this, organisations need to involve legal council to ensure regulators, auditors and other regulatory bodies are satisfied to the scope of the Information Security Policy. It is recommended that some form of information management system manage these policies.

3. For each information management system, organisations need to keep and maintain a *Procedures Manual* that includes: Data Capture, Data Protection (Backup and Recovery of both data and applications), High Availability and Application Performance, Security and Protection and Data Indexing (information document version control).

4. Organisations also need to ensure they have an *Infrastructure Manual*. This will describe, and show, how all the elements within the infrastructure, hardware, switch technology, software and network, interact with each other, detailing systems configurations and any changes made at any point during the system lifecycle.

5. Finally, of course, enterprises should provide an *audit trail* and report on all of the above. This needs to be sufficiently comprehensive to placate any court or external auditor and should include audit trails proving data and information usage as well as data integrity.

5.8   Storage recommendations for email

So, having adopted a business management policy that describes how long an organisation wishes to keep email records, IT needs to examine available options when considering how to store the data. Organisations do not want to store legacy data on the most expensive medium for 15 years or so. However,

1. The storage used needs to be secure enough to be able to maintain evidential weight in court.
2. The data needs to be able to be found and retrieved quickly and efficiently.
3. The data needs to be stored on the most appropriate media for the point in its lifecycle and frequency of retrieval.
4. Legacy data needs to be able to be read by future application versions or legacy application.
5. Versions need to be made available for retrieval; as well as the ability to automate deletion when the retention period expires.
6. Like-type data needs to be stored intelligently: deleting individual bytes of data from a tape could present problems. It makes more sense to store data with the same deletion requirements on the same tapes.

The answer to all these issues can be overcome by implementing an intelligent email archiving or indexing system. HSM, DLM or ILM systems are designed to capture data – in this case email records – in a structured way. The easiest way to achieve this is to have a dual system:

1. a journaling mailbox on the email server that receives a copy of all internal and external emails;
2. a version used by the employee.

All emails go into the archival system. Junk, spam and potential virus emails or attachments are removed from the end user's version, and end users can move, save and delete files and emails as they require in their normal business day-to-day activities. The archive holds *all* the emails to maintain records in their entirety in order to provide evidential weight. It is a carefully controlled store that typically allows read-access to the originator of a message if deletion happens by accident or to a privileged supervisor, but no deletion rights. Write or delete access is limited to administrators and any change they make to messages is recorded in an audit trail.

In order to exploit past email records for evidence in court or other business purposes, enterprises need to be able to find them, preferably without long painful searches through backup tapes that must first be restored to the mail system. It simply makes no sense to save data if you can't find it again when needed. So, when an email archival system processes a new message, it indexes the complete contents of both the body and the attachments before securely storing the message as an encrypted compressed file in the archive. The index is retained in the email archival system's database, allowing the user to perform a Boolean algebraic search of both the message body and its attachments using all the possible criteria at the search engine's disposal, including sender or recipient name, date, subject line description or even keywords taken from the text of the body of the email message.

Storing any kind of data can be a difficult decision. The storage media for an archive may be disk, tape, WORM drive (Write Once Read Many), or combinations of storage media depending on frequency of retrieval requirements, day-to-day access, or full archive needs. However, the only change email users see is that the normal icon is replaced with an alternative symbol for an archived email. This informs users that particular records now reside in an archive rather than the email server's message store. The retrieval time is relative to the media type storing the email archive. If it's near-line disk storage, then access may be nearly as quick as if it was held on the local machine's hard drive or the email server itself.

Having created an archive, how does an enterprise read that information in a few years? Or, what happens if someone decides to install a competitive email system? This could obviously cause issues reading old emails – backwards compatibility is not often maintained within one vendor through several iterations of development, let alone to a legacy competitor's system. It therefore makes absolute sense to maintain HTML format copies.

What happens when an email reaches the end of its retention period? Under normal circumstances the data, or email, would first have to be retrieved from a physical archive location (usually offsite), the dataset reloaded onto a compatible server and then the data or email record deleted and the whole archive copied back onto another tape. In the case of archiving when an email reaches the end of its retention period, the system flags it for deletion, either performing the deletion automatically or, more normally, requesting the IT admin to confirm the deletion first. Email archive applications allow IT managers to centralise corporate messaging data, not only for compliance with government and other mandates but also to keep the data within reach of knowledge workers. Some of the more sophisticated archiving tools allow IT managers to locate stray data more effectively and give mobile users continuous access to corporate archives via online and offline vaulting systems.

In addition to making records easier to find and more likely to stand up in court, email archival systems can also reduce an email server's message store size, helping avoid costly upgrades, minimising backup time and optimising performance and reliability. Many organisations implement email archival systems for this reason alone. Managing exchange .pst files can be an IT administrator's nightmare and migrating .pst files from one storage system (especially where the majority of .pst files are stored on the local desktop/laptop itself) to another can be an impossible task. Exchange can proliferate the disorderly sprawl of .pst files as a by-product of email quotas imposed by IT in order to contain the size of the messages stored on the Exchange servers. Endusers then combat the quotas by saving mail messages as .pst files and storing them on local hard drives or the network. This makes the .pst files nearly impossible to find by IT. In addition, .pst files are highly susceptible to corruption and accidental erasure.

Using an email archiving system effectively can become effortless because older messages and large attachments automatically migrate out of email in-boxes and onto an archive. This migration is virtually transparent to users. Migrated messages still show up in Outlook – they just have a small icon on them signifying that the messages have been archived. When users click on such messages, the data is retrieved as if it were on the user's email server. How can you achieve this Utopia? Finding and migrating .pst files already saved to users' systems may involve installing a .pst discovery and migration tool. Once these files migrate to the archive, either by push (where the end user can send files to the archive) or by pull (where the archive itself goes out and discovers the .pst files through the network), they can be protected and indexed for future reference.

5.9 Conclusion

There is no legal obligation to retain email records unless their particular nature renders them relevant to legislation or regulatory body requirements. But there is a legal requirement to keep certain types of records for a defined period of time. Each organisation must observe all general legislation and examine any appropriate sector-specific regulation. The potential requirement to prosecute or defend civil and criminal actions makes email retention a sensible precaution. It also makes sense to classify email records for the purposes of assigning separate retention periods.

Email archival systems offer a practical way to store emails in a manner that provides evidential weight and allows rapid retrieval of records for both legal and business purposes. Over 75% of discovery requests made by lawyers are now for email. The weight of these requests puts pressure on IT to manage the costs associated with retrieving archived email from backup tapes. Given the recent trend in heavier fines for nonretrieval of email, simply agreeing to pay a nonretention fine is now not likely to be an option. So email archiving not only improves ITs capabilities for retrieving email but also takes the management of potentially sensitive emails out of the end users' hands and puts it back in control of the IT department. Establishing business policies to improve retention and retrieval requirements, and providing users with some retrieval management capabilities, is the only way to manage email retention, deletion and retrieval.

6

Security

Corporate data and information is a 'multi-tentacled beast' and is becoming increasingly difficult to corral and protect.

John Mallery

6.1 Alerting organisations to threats

Information security is no longer the sole responsibility of System Administrators or even IT Managers. Since organisations now conduct a significant portion of, if not all, business through electronic means, today's increased cyber attacks mandate that organisations must ensure the data and information required to conduct business is safe and secure. This clearly implies a well-conceived data management, storage, retention and recovery strategy.

Security issues now typically involve board level oversight. It therefore follows that all organisations are well-advised to consider information security seriously as an integral element of their data lifecycle strategies. A few years ago, the largest threats, simple viruses, were probably nuisances rather than threats. However, yesterday's nuisances are now major traumatic cyber attacks.

The situation is increasingly grave. Most attacks are now blended threats (also known as 'blended attack', 'combined attacks', or 'mixed techniques') aimed at specific organisations rather than email or general Internet attacks. Indeed, there are

Data Lifecycles: Managing Data for Strategic Advantage   Roger Reid, Gareth Fraser-King and W. David Schwaderer   © 2007 VERITAS Software Corporation. All rights reserved.

a number of fundamental challenges for the IT industry and enterprise organisations around the world.

The world we live in is reliant on IT systems. Any interruptions are unacceptable to businesses. However, interruptions are inevitable – new viruses are appearing, attacks are being planned and power issues are disasters just waiting to happen. So, the question remains: how do organisations stop the onslaught that has already been launched or has already happened?

As organisations strive to grow and prosper, they need secure and reliable availability of IT services to keep the business operational at all times. And that requires building an infrastructure that is flexible enough to respond to a changing IT environment, but resilient enough to withstand disruptions. IT is changing. It is clear that the change is driven by the need to align IT with the business and move away from the traditional cost-centre focus IT has concerned itself with. Now, IT needs to purchase devices and components as a business-aligned service oriented provider.

Thirty years ago, mainframes supported businesses – it was utterly controllable but somewhat inflexible. Open systems gave a business the ability to manage its requirements more efficiently, but promoted a growing complexity of information technology and IT infrastructures. IT countered new demands by installing additional servers and storage capacity to counter storage and data bottleneck problems. Newly provisioned extra capacity was usually dedicated to specific servers. The inevitable result was that more staff were required to administer the increased capacity. Consequently, the infrastructure became disjointed and increasingly difficult to manage. The IT industry is now moving towards the need for a more intelligent way to manage both the business requirement and the IT infrastructure.

However, critical tools aren't always interoperable. Moreover, since IT operations and security functions often have conflicting priorities, any solution can introduce more complexity and more problems than they solve. That's why organisations need a holistic approach to create balanced and interoperable solutions. It's an approach that helps align the right people, processes and technologies required to build resiliency into the entire infrastructure – from storage, critical servers, essential applications and the network gateway, down to individual clients. It's an approach that improves the business value that IT brings to organisations.

In the past, organisations have balanced two conflicting goals: providing easy information access and securing information against malicious threats. Organisations need the ability to provide both requirements without having to compromise either of them. They need to take a smarter approach to building and managing their assets and organisations. Companies now must:

- Set a baseline by evaluating their assets and security systems, assessing current procedures against risks and setting goals and policies for desired levels of availability and security.
- Identify and analyse threats, creating plans to remediate them.
- Deploy proactive safeguards against potential threats, enabling them to protect assets and recover from disruptions quickly.
- Remediate threats quickly and implement long-term solutions for vulnerabilities, including regularly updated policies and new certification standards for compliance.
- Proactively monitor their security, information management and storage systems to provide a clear view of the entire infrastructure and allow a rapid response to any disruption.

Storage and security remain at the forefront of IT departments' minds. Data, information and the means to act upon that information (applications) remain among the most significant business assets. Instead of treating IT as a business by-product, organisations must carefully determine how to protect their IT infrastructure, and how to access and share data, and ensure its availability to current and future applications. Managing information comprises significantly more than simply putting another storage device on another server and ensuring it is protected from viruses.

As critical components of every business application, both security and storage need the same careful, robust architectural design as networks and systems. The IT architecture is more than simply the topology of how to connect clients to applications, databases, servers and storage, and how to protect the infrastructure against malicious threat. IT architecture must include the people, processes, hardware and software that support data in the organisation. To face the challenges of a changing business environment, organisations need nimble architectures that adapt to change quickly and efficiently.

IT infrastructure Management is all about

- supporting day-to-day, week-to-week, month-to-month peaks and troughs of the business environment;
- the ability to grow or shrink business requirements;
- adapting to change;
- building new routes to market;
- innovating;
- retaining customers and improving the customer experience;
- managing the processes required by the business;
- managing applications and data intelligently;
- ensuring continuity of operations;
- restoring applications quickly and efficiently should a disaster strike;
- constantly monitoring and providing the right applications to the right people when they are required at the right level of service.

Basically, average business individuals do not care what the IT infrastructure does or how it does it. They just want to access applications and data with the knowledge that the infrastructure performs in the manner required and that the organisation and the individual is not threatened by attack from viruses, phishing, corruption, stealth or any other malicious attack. In short, they want to know they are safe and they want to conduct their activities without fear.

To this end, organisations seek the broadest range of solutions to establish processes that increase the efficiencies of IT security and operations teams and align them toward the same goal: to build an infrastructure that ensures the security and availability of information systems and preserves business continuity. The consequent security, availability and storage solutions hopefully offer customers everything that organisations need to build a foundation for enduring success and growth.

Organisations must be able to:

- secure their environment; not just throw everything behind a firewall, but secure every single end-point deployed;
- discover assets they have in their IT department;

- re-architect and consolidate the environment to gain efficiencies in administrator productivity and resource utilisation;
- standardise – classify applications and agree upon specific vendors for necessary requirements to run them;
- manage the entire environment using a standard set of tools so application service levels become measured and predictable;
- automate processes to drive down cost, improve application service levels and make IT a more predictable organisation;
- simplify the updating of *all* the infrastructure through patch management and upgrade management capabilities;
- manage and control the entire environment efficiently and cost effectively – in ways that adds value to the business;
- develop internal security policies which discuss how storage and data integrity products will be used in the environment and highlight how the products could be potentially misused;
- insure that data products have authentication and authorisation services that validate identity and protect communication channels for message integrity.

6.1.1 Vulnerability identified and early warnings

Historically, 'Research and Development' groups developed products, 'Operations' manufactured the goods and 'Sales' sold them; but not anymore. CIOs are now not only faced with enterprise resource management and e-Business requirements, but also security and availability convergence. One can see the need for IT integration falling onto the CIO's desk and, indeed, where else would the board turn to with enterprise application integration? But convergence of the business and IT refers to much more than just these added responsibilities. Now CIOs are concerned with the way the business uses and protects the information held within applications – the data held so dear – together with data business use and integration with other systems. This is beginning of the data integrity age.

Data vulnerability identification of enterprise systems and legacy applications will soon become a top priority of senior managers. Identification of business critical applications and servers that are needed to be protected should be the first step in this process. Staff resources and financial capital, together with information (data), are now of the most important assets of any company. Therefore, once the systems and applications are identified the next step is to identify the vulnerability, remedies and required patches. There are several public sources that one could use to capture data about vulnerabilities; some of these sources are listed below:

• Help Net Security: http://www.net-security.org/vuln_main.php
• Bugtraq: http://www.ntbugtraq.com
• Full-Disclosure: http://seclists.org/lists/fulldisclosure

6.1.2 Early awareness of vulnerabilities and threats in the wild

As if CIOs didn't have a huge responsibility before, they certainly do now and this should be recognised accordingly. CIOs now manage the entire responsibility for the organisation's information assets; they decide on hardware and software architectures as well as on the relevant security policies; they set service level agreements, key to any disaster recovery strategy, and are responsible for prioritising new applications (which can mean the difference between success and failure of the company). But, unlike most managers, CIOs can't have an off-day. They can't assume they can relax for even a minute because, unlike most managers, they proceed under perpetual scrutiny – when a system slows, the entire organisation knows it.

Although the industry has vulnerability lists as discussed above, it should be known that some companies fund lists and could be biased in one form or another on certain vulnerabilities. Don't let that keep you from being proactive in your security goals, but do filter a majority of the rhetoric. Be very wise in how you apply these vulnerability lists and don't always use the same contact names.

Because it has changed, organisations must necessarily understand the diversity of the CIO's role. It is still predominately regarded as application implementation rather than an effect of decisions that clearly affect information as a resource and, as a result, a company's business. IT security strategy requires the long-term and short-term view. With the need for information to become more integrated within the business, the need for the

CIO to evolve from an IT manager to a general business security strategist/manager is essential.

It appears that, though most senior managers do give some thought to Enterprise Resource Planning (ERP), e-Business software and storage, they have not considered the need to integrate the applications and storage with security in mind. This is a major stumbling block to say the least. Without upper and mid-level support for a security policy, the security efforts will fail and efforts to protect the data and its information will fail as well. One security book (Mallery *et al.*, 2005) eloquently states; 'The actual burden is on the technology administrator and analyst to apply business principles to justify the improvement of the security and protection of the critical business information.'

We all know that Return on Investment (ROI) is the critical metric in any investment decision, especially in the current economic climate. The expected benefits of a proposed project must exceed that project's anticipated costs by enough to justify the inherent risks. Again, this must now be achieved in a relatively short order. With IT hardware shelf life becoming increasingly shorter, it is commensurately harder to write off any IT spent over a five or even three year period. So, ROI needs to be validated within six months to a year rather than over several years.

Because of the current financial climate, finite capital resources may still shelve a project if other investments promise to deliver better returns. A proposed managed data protection and security project will only proceed if a rational comparison of the projected costs and benefits can justify the capital spent.

Most IT application project benefits are a direct result of new application functionality. Most applications are designed to increase revenues, while systems that automate functions, and paper-intensive processes, reduce costs. The benefit logic is reversed, however, for a managed Data Protection and Security project, since the focus is not on what is gained by implementing the project, but what is lost by not implementing it. And this is what CIOs consider: not what is gained, but what is potentially lost if the project does not proceed. Therefore, missing or ignoring costs when conducting a cost/benefit analysis for a Data Protection project often results in a gross underestimation of the project's potential benefits.

As always, the bottom line is profitability. Every business should manage data within regulatory requirements to the most appropriate level for its unique operational, marketing, competitive and financial situation. Finding the optimal level of data management investment requires considerable effort. But taking the time to investigate and understand the interdependencies certainly helps to quantify any potential downtime impact better. This is an essential first step in the quest for a data management programme.

A majority of the upper management usually leave such security and technical subjects to their IT department, which tries to manage this costly task essentially without board intervention. By concentrating on business critical applications, companies lose sight of the safety of the information stored within them. Furthermore, data security is discussed in isolation at departmental and application levels, creating still more information silos throughout the organisation, defeating the primary goal by permitting users to access all existing application information without the company's knowledge.

6.1.3   Listening posts

Given their relevance to strategic decisions for senior management, data integration and protection now involve much more than simply mastering security at the lowest levels. In the final analysis, it involves the secure integration of information or content stored in diverse applications. The management and staff must recognise security as of strategic value to the company in order to recognise one of the company's most important assets – information. Therefore, it is the entire company's task to avoid making unlimited access to information the goal of every IT-related decision. Rather, the direction should be towards limited access to information that is only of use for legitimate reasons.

Staff members should be aware of forthcoming attacks, always listening for new news and preparing for the worst. Having a proactive strategy helps the enterprise stay competitive and in business. For those who want to be proactive, consider the next set of 'IT Security Listening Posts' pointers:

• US-CERT: http://www.us-cert.gov
• Common Vulnerabilities and Exposures: http://cve.mitre.org

Businesses must manage and secure their growing data demand as well as increasing data volumes. Data's manifestation and safety is one of relentless storage complexity and escalating costs. The complexity comes from an endless number of proprietary, server-siloed storage islands, each with unique quirks, proprietary management tools and security holes. Some devices are bursting at the seams, unable to consume another storage megabyte, but harbouring yawning security holes, while other resources remain idle with storage to spare, unable to share it with applications and users, but safe from external intruders.

The exigent total cost comes not from the physical storage unit itself, but from the cost of managing and securing the storage infrastructure, which has ballooned to five times the physical cost per megabyte. In fact, research firm Gartner estimates that only 20% of storage costs come from the storage hardware itself – administration accounts for nearly 15%, backup and recovery 30% and downtime 20%. And, what about security costs?

The storage nirvana is an ability to safely control and manage all storage, the data lifecycle and information created by the business from a central location, using a standard set of storage management tools and security policies across the entire storage network. The storage solution eliminates storage complexity and allocates storage, as appropriate, from one giant storage pool while providing for data integrity.

The idea is to virtualise disk resources by taking multiple physical storage devices and combining them into logical storage devices that are allocated at will to applications and users in order to ease the management of complex storage environments. But with an open security policy that is endorsed across the company (such a scheme comprises the ideal storage environment); one that provides the ability to allocate secured storage capacity through requirements for read/write performance, the type of storage availability, or how much storage is needed. Remember, ill-performing applications could mean they have been compromised.

6.2   Protecting data and IT systems

E-Business on the Internet has created an important business channel. It allows businesses to offer products and services

24 hours a day, 7 days a week with relatively small resources. Previously, this was only within the reach of large enterprise organisations. Now, companies of any size can offer customers their wares through the Internet with a small investment. This form of business is appealing to customers who do not want to spend time buying products and services the traditional way, or to customers who are not within the immediate geographic area of where the products are offered. The Internet has become a perfectly valid and vital distribution channel.

Personalisation of Internet services can potentially create unique customer relationships. As various sales strategies have improved online shopping experiences, organisations are under increasing pressure to provide these services and systems as fast as possible. The systems that run these Internet stores must be accurate and available at all times. These services must be easy to use but absolutely secure, especially with the use of credit card payment and personal information being held on a server that is effectively online to anyone. In addition, services are expected to be online 24 hours a day, 7 days a week. E-Business customer requirements place huge demands on the IT department in terms of keeping the online retail machine up and running and secure, as well as dealing with the inevitable management of the life of the personal data held by the organisation.

6.2.1 Threats blocked using vulnerability signatures to prevent propagation

Malware (malicious code) security incidents have been steadily growing over the last few years. A few definitions are useful:

- A computer virus is a piece of malicious code that attaches itself to, or infects, executable programmes.
- A Trojan horse is a malicious programme disguised as something else, something harmless, that relies on the end user launching the code which then deletes or steals data.
- A worm is a programme or algorithm that replicates itself over a computer network and usually performs malicious actions, such as using up the computer's resources and possibly shutting the system down.

Current threats are causing multiple disruptions across the Internet. Great places to visit to start researching these threats and gather vulnerability data are

- National Vulnerability Database: http://nvd.nist.gov/
- Open Source Vulnerability DataBase: http://www.osvdb.org/
- Common Vulnerability and Exposures: http://cve.mitre.org/

The first item (NVD) is a comprehensive cyber security vulnerability database that integrates all publicly available US Government vulnerability resources and provides references to industry resources. It is based on, and synchronised with, the third item, (CVE) vulnerability naming standard. The NVD database includes computer vulnerabilities at the OS, application or firmware level. The CVE can also quickly provide an overview of the terms.

6.2.2 Preventing and detecting attacks

Once organisations understand what assets they need to protect, and where the vulnerabilities are, then are in a much better position to stop the threat or attack from ever happening. As this book is aimed at managing stored data we have confined our approach to data assets rather than the bigger cyber security picture. The list below is not exhaustive, but it does give a start on stopping threats and attacks:

- encryption;
- authentication and authorization.

Without encryption, losing tapes can be crippling to a business. In a recent example, a large United States financial institution misplaced data tapes on approximately 3.9 million customers while shipping them to a Texas credit bureau. If those tape had fallen into the wrong hands the personal information held on the tapes would have been crippling for both the company and the persons whose information was lost or stolen. This may seem as a bizarre example but stranger things have happened. What saved the company and the individuals was the fact that the data on the tapes was encrypted and therefore useless to anyone trying to use the information for financial gain or corruption.

Even though we don't intend to discuss encryption intricacies in this book, we do mention areas that would help prepare data centres for attacks or data loss, and encryption is one such area worth examining. Commercial hardware and software solutions are presently available and primarily use asymmetric or public key cryptography. Asymmetric cryptography is where a pair of keys is used to encrypt and decrypt a piece of data. Asymmetric cryptography is usually associated with the sending of messages in packets so that the message arrives safely and securely. This process is known as a public key infrastructure. Since such encryption requires one key to encrypt data and a different key to decrypt, encryption does involve system performance overhead. But once again, is what the data encryption protects more valuable than the cost to protect it? Any company's data protection strategy should include encryption requirements and establish the business requirements, as well as the legal, regulatory aspects.

Authentication is also a major consideration in the quest to protect information assets – that is, authenticating application access and application data resources. One primary authentication mechanism is passwords. Yes, this is well known, but probably not enforced as often as warranted.

Using passwords is increasingly becoming standard. However, authentication and encryption go hand and hand. Therefore, it is essential to understand the environment in more detail by first reviewing it and then ensuring encryption and authentication work effectively.

6.2.3 Managing security in a data centre

Enterprise information is created, stored and used by various users who expect data accuracy and instant availability. Information stored can range from financial, legal and technical materials to customer-related data. The need to ensure secure access, usage and storage of information is of critical importance since business downtimes or compromised information can easily cripple a business. There is one critical challenge with NAS/SAN solutions – given the permeable network with its incessant vulnerability to network layer attacks, are second generation storage solutions truly secure enough?

6.2.4 Monitoring and identification of systems versus vulnerabilities and policies

A fundamental information security goal is to protect the confidentiality, integrity and availability of information assets. Activities that adversely impact these information attributes are referred to as threats. Threats occur because of information associated system vulnerabilities. Risk is directly proportional to the level of an information asset threat. No single networked storage device comprehensively secures networked storage and leverages all existing security functions.

First generation server attached storage systems (DAS) ran into size, scalability and manageability limitations, and the unexpected challenge of an increasingly networked world. The second generation networked storage solutions (NAS/SAN) leveraged IP-based investments and were created primarily to overcome DAS limitations of manageability and accessibility. However security on these systems was overlooked, opening them up to increased security risks. Current security measures do not provide effective defence to the new security challenges posed by networked storage and require businesses to re-evaluate their security strategy.

Today, NAS/SAN devices can and do provide a distributed, resilient and stronger protection level compared to a centralised single-device security model with a concentrated vulnerability point. While the benefits of storage networks have been widely acknowledged, enterprise data consolidation on networked storage can pose significant security risks. Hackers adept at exploiting network-layer vulnerabilities can now explore deeper corporate information strata.

Key drivers to implementing effective networked storage security include:

1. *Perimeter defence strategies* that focus on external threat protection. However, with the number of security attacks increasing, relying on perimeter defence alone does not sufficiently protect enterprise data, and any single security breach can cripple a business.

2. *The number of internal attacks is increasing*, thereby threatening NAS/SAN deployments that are part of 'trusted' corporate networks.

3. Certain industry verticals must conform to *compliance regulations*.

4. *Market globalisation* demands corporations and their partners meet security directives in all parts of the world – implying that all distributed storage needs to be secure.

Compromised storage risks range from tangible losses present with business discontinuities in the form of information downtime, to intangibles, such as the loss of stature as a secure business partner. With the number of reported security attacks increasing, a firm understanding of networked storage solutions is a precursor to determining and mitigating security risks. However, storage security issues are often overlooked because few IT professionals understand both security and storage. An effective networked storage security policy can be built only with a clear understanding of networked storage technologies.

6.2.5   Responding to threats and replicating across the infrastructure

This chapter has underscored the need for data security in information support infrastructures. Even if an organisation doesn't follow best security practices, it would be nice to believe its data is safe. Forget it. The bad news is that if security doesn't 'measure up', it probably doesn't provide an effective way to respond to the plethora of existing threats.

Be aware that threats normally morph into attacks and could appear in the form of

- spyware;
- viruses/worms;
- phishing (fraudently obtaining information);
- zero-day exploits.

Spyware is any technology that aids in gathering information about a person or organisation without their knowledge. On the Internet (where it is sometimes called a *spybot* or *tracking software*), spyware is programming that is put in someone's computer to gather information secretly about the user and relay it to advertisers or other interested parties. Spyware can get into a computer as a software *virus* or as the result of installing a new program. There are software applications that aid in detecting and removing spyware and these should be implemented throughout the company to avoid exploitation of corporate data.

Today, spam is recognised as public enemy number one. Not only is spam annoying and time consuming, users must also rid their email inboxes of this junk. Yes, it's annoying, but it also prevents employees from being productive because they have to spend so much time removing it.

There are many different ways to stop spam. A first starting point could be using a client based anti-spam system to stop it or employ a gateway approach. Each of these anti-spam approaches are available from a variety vendors and can be found on Symantec and FrontBridge Technologies web sites.

Phishing is becoming increasingly difficult to defend against because of the sophistication of the expeditions. Application layer firewalls are a good source of protection. But always remember, be suspicious of online information requests and understand their source. For more phishing information, the Federal Trade Commission (FTC) enables web surfers to report email scams, like phishing at http://www.ftc.gov.

A zero-day exploit is one that takes advantage of a security vulnerability on the same day that the vulnerability becomes generally known, so that the vulnerability and the exploit become apparent on the same day. Since the vulnerability isn't known in advance, there is no way to guard against the exploit before it happens. Therefore, companies exposed to such exploits should institute procedures for early detection of an exploit.

Ultimately, enterprise requirements drive any decision for any selected solution. However, enterprises should carefully consider the advantages and disadvantages of all candidates before selecting a product.

6.2.6 Patches and updates implemented across infrastructure

Security patching is an extremely important activity for system administrators within enterprises, which depend upon them to properly patch vulnerable systems from either malicious code or known attacks. This process is on-going and should never stop. A company should employ a routine process to properly patch and prepare their systems against attacks. A key requirement of patching is that administrators know what systems to patch. This is often an enormously big task, but must done. Here are some basic considerations to help develop a patch strategy:

- know what systems should be patched;
- know where to obtain the patches;
- understand how the patches are verified with patch validation;
- test the security patch prior to production rollout;
- design test plans with its associated patch back-out plans;
- document all patches and implement change management.

6.2.7 Keeping information secure and available

Remember, building an effective networked storage security policy requires a clear understanding of networked storage technologies. It is imperative to understand the environment and the technologies comprising it. There are many 'moving' parts in the environment with different vendor's hardware and software. Once the systems needing protection are identified, locate their vulnerabilities, seek the appropriate remedies and clearly understand probable and describable threats.

In the end, document security and procedures maintain management buy-in. As mentioned earlier, any security strategy should produce profits that are visible to the business. Needless to say, it is important to account for the time required to implement a security policy, and time and resources are both needed to be proactive in protecting information assets. Prevention is what keeps enterprise data available and ultimately accessible to customers.

6.3 Conclusions

What are the concerns IT departments have about possible security threats to the enterprise? Security issues have been around for many years now. Fundamentally made up of viruses, hackers, spam, identity theft and denial of service, security issues are on the rise. Now, threats are no longer centred on virus writers and hackers attacking enterprises simply because they can, but are designed to attack enterprises and individuals for personal and corporate gain. Exploits, vulnerabilities and buffer-overflow techniques (introducing viruses and worms to a PC and then helping to spread the virus) have been used by malicious hackers and virus writers for a long time. However, until recently, these techniques were not commonplace in computer viruses. The CodeRed worm was a major shock to the security industry as it was the first worm that spread not as a file, but solely in memory by utilising a buffer overflow in Microsoft IIS. Many antivirus companies were unable to provide protection against CodeRed; only companies with a wider security focus were able to provide solutions to the relief of end users.

Many other similarly successful worms followed CodeRed, such as Nimda and Badtrans, as new techniques are often picked up and used by copycat virus writers. Security exploits, commonly used by malicious hackers, are being 'blended' with computer viruses resulting in very complex attacks, which in some cases go beyond the general scope of antivirus software.

A blended threat is where the virus exploits some sort of security flaw of a system or application in order to invade new systems. A blended threat exploits one or more vulnerabilities, but may also perform additional network attacks such as a denial of service against other systems.

In the past, a large gap existed between computer security companies, who focused on both intrusion detection and firewall techniques, and the antivirus companies. Security and viruses were almost seen as two separate issues and some computer security professionals did not consider computer viruses seriously as part of security, or they ignored the relationship between computer security and computer viruses. When the CodeRed worm appeared, there was obvious confusion about which genre of computer security vendors could prevent, detect, and stop the worm. Some antivirus researchers argued that there was nothing that they could do about CodeRed, while others tried to solve the problem with various sets of security techniques, software and detection tools to support their customers' needs. This shows that organisations need a holistic approach to their security strategy covering: intrusion detection/prevention, antivirus, anti-spam, system scanning, firewall/gateway technologies, updating and security alerting services, as well as security management. All these elements are required to combat the modern IT security threats that now threaten the enterprise.

With the advent of CodeRed the security industry was simply turned on its head: the intermediate solutions were criticised by antivirus researchers, instead of realising that the affected customers needed a number of security tools in order to combat these new threats. IT security suddenly became far more complex, with single solutions simply not being good enough to combat the new blended threats. Now, not only are antivirus patches required, but also further requirements for cleaning, deleting and intrusion prevention, and the management of installing the many thousands of patches required in a large enterprise environment. This process is never easy and enterprise IT departments fear that new patches could introduce a new set of problems, compromising system stability. Blended threats now require that we provide multi-layered security solutions that can be delivered in a combined effort to deal with these attacks.

Traditional antivirus products do not scan Windows memory, which means that blended threats often remain, undetected, in the memory of the PC. Some intrusion detection products simply alert rather than block when a blended threat is located. So all the technologies are needed to protect the enterprise: firewall and gateway security, antivirus, spam intrusion detection and prevention and updating techniques, so the threat is blocked when detected, removed from the system, the security policies are updated and the network, servers, PCs and any caches (memory) are scanned for the threat signature.

On-demand memory scanning solutions are still unable to prevent the worm from entering the system. Active protection for such an attack needs to be identified elsewhere. Firewall software can be configured to prevent certain types of attacks from happening in the first place. However, firewall software will rarely prevent against all new kinds of threats and will require expertise in dynamic maintenance of firewall configuration. Using a firewall to mitigate the risk is not limited to the standard enterprise firewall, but also includes personal firewall. The combination of personal and enterprise firewall can be extremely powerful in dealing with not only incoming but also outbound attacks. This is especially important with the number of enterprises with mobile workers who use laptops, PDAs, Blackberrys and mobile phones that they synchronise with the main enterprise network. Half of the damage is done when the attack enters the internal network; the other half occurs when it leaves the internal network. The outbound damage can often be more costly than the inbound.

Reference

Malleny, J., Zann, J., Kelly, P. and McMullin, R. (2005) *Hardening Network Security: Bulletproof your systems before you are hacked.* McGraw-Hill.

7

Data Lifecycles and Tiered Storage Architectures

7.1 Tiered storage defined

At the turn of the millennia, storage subsystems generally consisted of two *types* or *levels* storage – *online* and *offline* storage.

- High-performance and mission critical enterprise applications requiring fast user response times necessarily accessed online storage subsystems. These subsystems typically contained arrays of low-capacity, expensive, parallel Small Computer System Interface (SCSI) or Fibre Channel hard disks referred to as *enterprise-class* storage devices.

- To archive online data, enterprises used on offline storage subsystems that typically contained extremely cost-effective media such as tape cartridges or CD-ROM Write Once, Read Many (WORM) drives.

For decades, the storage industry diligently toiled to populate the vast continuum between these two storage subsystem extremes. However, the only potentially cost-competitive alternative storage device candidates were Parallel ATA disk drives (PATA) that repeatedly proved unsuitable. This undesirability persisted even

Data Lifecycles: Managing Data for Strategic Advantage   Roger Reid, Gareth Fraser-King and W. David Schwaderer   © 2007 VERITAS Software Corporation. All rights reserved.

after PATA disk drives evolved into Ultra DMA (Ultra DMA or UDMA) disk drives.

The unsuitability was a direct result of PATA/UDMA devices, referred to as *desktop-class* storage devices, generally being too unreliable under sustained enterprise application usage stress. Moreover, from a physical packaging perspective PATA and its UDMA successor devices also required cumbersome parallel signal cables and bulky connectors that complicated large array configuration and inevitable maintenance efforts.

Finally, although parallel SCSI or Fibre Channel hard disks required substantially more technical sophistication to configure and maintain, they generally adhered to industry standards. Without making the point too strongly, enterprise-class storage devices exhibited substantially superior interoperability and co-existence characteristics without exhibiting the frustrating hard-drive vendor master/slave device interoperability problems rampant among desk-top storage devices.

However, with inexorable hard disk areal density (also called *bit density*, which is the amount of data that can be packed onto a storage medium and is often measured in gigabits per square inch – it is used to compare different types of media, magnetic disks and optical disks), recording density advances, Serial ATA hard disk reliability improvements and widespread availability of Serial ATA hard disks providing massive storage capacities, storage subsystem vendors are now able to address the challenges associated with exploiting low-cost, desktop-class hard disks within enterprise applications. The result is a new type, or level, of storage subsystem, commonly referred to as *near-line* storage.

Near-line storage, then, is a hard-disk-based, *mezzanine* storage resource residing within the traditional storage hierarchy expanse between online and offline storage.

Thus, *tiered storage* is a storage approach that involves

1. online storage – uses enterprise-class devices;
2. near-line storage – uses desktop-class devices;
3. off-line storage – uses archival devices.

Because near-line storage access time, reliability and cost/GByte can easily range between those for online and offline storage alternatives, near-line storage can be viewed as providing a partial *storage solution continuum* between the online and offline storage extremes.

Today, the storage industry can provide a rich spectrum for storing digital data, each solution providing its own performance, availability and cost characteristics. Storage administrators can also minimise capital costs by choosing disk drives attached directly to server I/O buses. Or, they can select elaborate disk arrays that provide packaging, power and cooling, and centralised management for dozens or even hundreds of disk drives.

The missing link that had historically impeded this flexibility was Serial ATA storage devices. So, we now turn to near-line's fundamental ingredient: Serial ATA technology.

7.1.1 Serial ATA background

The movement toward tiered storage began inconspicuously at the Palm Springs, California Intel Developer Conference in March 1999. Intel executive Pat Gelsinger obliquely mentioned a previously undisclosed internal Intel research effort during his keynote presentation. He called it *Future ATA*. Specifically, Mr. Gelsinger stated:

> ... And beginning in the second half of 2000, we expect the transition to Future ATA. Again, following a very similar protocol architecture, move to a narrower, higher speed version of that, narrower for cost savings reasons and higher performance or higher speed for performance increases. We expect future ATA to begin in the second half of next year and to provide us the opportunity to increase performance in the I/O subsystem through 2005 and beyond.

When Mr. Gelsinger spoke, he also indicated that USB 2.0's projected bandwidth would range from 15.6 to 31 MBytes/sec. However, in October 1999, Intel announced USB 2.0's bandwidth would range between 40 and 60 MBytes/sec, an increased factor of three to four. The distinct possibility was that, in addition to absorbing external peripheral connectivity opportunities, USB 2.0 technology would also play a complementary role in accelerating the 'serial UDMA' follow-on (*Future ATA*) timetable.

Here, an obvious potential for a *discontinuous* change was the direct consequence of the USB 2.0 transceiver's new performance level. This new performance threshold clearly signalled that the low-cost, impending USB 2.0 serial *transceiver* could easily be a serious threat to SCSI mainstream hard disk attachment strategies.

Correctly implemented, and hence devoid of USB's protocol and topology limitations, a disk product using the USB 2.0 serial transceiver might easily provide the missing ingredient for industrial strength I/O subsystems that would generate extensive competition for SCSI across all its existing markets. This solution could clearly disrupt and perhaps even displace the enterprise storage device incumbent – parallel SCSI hard disks.

7.1.2 Serial ATA overview

On 15 February 2000, Intel, IBM, Maxtor, Seagate Technology, Quantum and two other partners revealed a new hard disk host connectivity solution named *Serial ATA* (or SATA). This new technology would supersede the UDMA's parallel signal interface methodology. To many, the transition appeared to be a minor, perhaps even superficial: but, it wasn't.

To understand Serial ATA's potential, it is useful to examine its key features.

- **Serial Signalling** – Serial Signalling allows Serial ATA to reduce the required number of signals from the 26 signals parallel UDMA used to 4, thereby reducing chip pin-out and system costs.

- **Software Compatible** – To system PATA/UDMA software device drivers, a Serial ATA device is indistinguishable from a legacy PATA/UDMA device.

- **Serial Cable** – Serial ATA devices connect to systems using inexpensive cables with compact connectors that are compatible with high-density server requirements.

- **Single Device per Cable** – Serial ATA abandons the PATA/ UDMA Master/Slave concept, only allowing one device per cable that a system views as a Master ATA device. This completely eliminates master/slave device interoperability problems while simplifying configuration and maintenance operations.

- **Serial Transmission** – Serial ATA uses 8B/10B serial transmission to transfer data over the serial cable. This high data-integrity transmission approach is widely regarded as *the* reigning *de facto* serial transmission scheme. Gigabit Ethernet, Fibre Channel and PCI Express also use this transmission scheme, as do many other technologies. 8B/10B is vastly superior to Parallel PATA/UDMA's parity checking which only checked data transfer operations using a weaker checking methodology known as *parity*. Moreover, because PATA/UDMA task-file register command transfers lacked checking, devices could receive *write* commands that were actually transmitted as *read* commands, resulting in storage corruption when random data was written to PATA/UDMA devices.

- **Low Voltage Differential Signalling** – Serial ATA uses low voltage differential signalling (LVD) with 250 mV offset that is compatible with both existing as well as emerging circuitry. It is also consistent with low power and cooling requirements.

- **10 Year Growth Road Map** – At Serial ATA's announcement, the development partners planned three generations that would transmit at 1.5 GBit/sec, 3.0 GBit/sec and 6.0 GBit/sec respectively. This enables respective 150 Mbytes/sec, 300 Mbytes/sec and 600 Mbytes/sec device burst rates with over 93 % efficiency maximum throughput.

The collective advantages the above features introduced meant that Serial ATA's initial announcement posed clear, present and formidable challenges to both parallel SCSI and the latter eventually became Serial Attached SCSI technology (SAS). Indeed, SAS eventually found it necessary to extend SATA device support despite strong initial resistance.

Today, correctly implemented near-line Serial ATA storage systems can use a multiplicity of inexpensive compact connectors and cables to provide high bandwidth I/O subsystems using inexpensive desktop-class devices providing massive capacity. However, it is fair to say that not all initial Serial ATA aspirations have been completely achieved at this time. For example, Serial ATA connectors are only rated for a few insertions/removals and internal Serial ATA cables presently tend to be very fragile – internally breaking if repeatedly flexed.

Originally conceived as a long-term solution to the high signalling voltages required and increasingly problematic UMDA parallel cables/connectors, Serial ATA's enterprise-class applications were far from assured in the eyes of some. However, Serial ATA's onslaught is in full swing as evidenced by Gartner's projection that Serial ATA will represent 94.3 % of total desktop and mobile PC hard disk drive shipments by 2005. Serial ATA now holds a commanding advantage in desktop systems, entry-level servers, and entry-level NAS systems. The advantage SCSI once enjoyed over its competition appears significantly reversed with Serial ATA holding enviable advantage in many dimensions.

Ironically, in the eyes of many, the adoption of Serial Attached SCSI (SAS) hard disks is now in question. Caught from above by high-reliability Fibre Channel enterprise hard disks (that are enjoying traditional cost reduction trends) and from below by Serial ATA hard disks (that are experiencing wide-spread adoption, increasingly reliability, application growth and feature enhancements), SAS hard-disk offerings face a difficult road as their supporters begin to navigate SAS's disruptive early deployment and always-painful technology stabilisation phase.

Indeed, many feel that if SAS had not eventually provided Serial ATA support, it could have failed to garner any market support *despite* the enormous support system OEMs and disk industry afforded it. As it is, SAS's role might increasingly appear to be highly marginalised, eventually proving to be a Fibre Channel storage-fabric replacement, only connecting hosts to Serial ATA drives. These are truly exciting times.

7.1.3 Serial ATA reliability

A primary remaining obstacle to enterprises deploying Serial ATA storage devices in mission-critical applications is the relatively lower reliability that Serial ATA hard-drives usually have as compared to enterprise-class hard-drives. However, it is essential to note that a disk drive's reliability is *completely* independent of the drive's host interface. That is, there are enterprise-class Serial ATA hard disk drives as well as desktop-class Serial ATA drives. It is also true that some manufacturers are offering enterprise class drives with SATA interfaces – a trend that may well continue and gain momentum eventhough such enterprise-class SATA drives presently cost more than desktop drives and can have noticeably lower storage capacities.

That noted, it is generally true that drives described as *enterprise-class* exhibit higher reliability over *desktop-class* (e.g. non-enterprise) drives. So, when investigating the matter, it is not unusual to learn that enterprise hard disk vendors often build and dedicate manufacturing assembly lines for enterprise hard disk manufacturing. On such lines, the soldering tolerances are stricter, component specification tolerances are tighter and the components have higher reliability as well as harsher duty cycle requirements when compared to their desktop counterparts.

For example, desktop drives can use lower-cost bearings that generate more vibration and wear faster than enterprise drives. The consequent vibrations present in large disk drive arrays, coupled with the lower accuracy actuator seek operations that desktop disks often employ, can cause significant operational and performance problems. Indeed, some Symantec storage testing labs vibrate their buildings. So, vibration and tolerance is a genuine concern.

These realities are partially reflected in two hard disk metrics – the Bit Error Rate (BER) and its Mean Time Before Failure (MTBF).

7.1.4 Bit error rate (BER)

A BER value indicates the average number of bits transferred before an error is detected and there are two different BER values associated with a hard disk model. At the media transfer read interface, the associated BER is the *raw* BER. Here, transferred bits are the encoded bit values detected on the media surface by the read/write head. Nominal raw BER values are 10^{-5} to 10^{-6}, meaning that one bit in 100 000 to 1 000 000 are incorrectly detected. Here, if 10^{-N} is good, then $10^{-(N+1)}$ is better. So, a BER of 10^{-6} is more desirable than a BER of 10^{-5}.

Disk drives typically use Error Correction Codes (ECC) provisions in an attempt to detect and compensate for these raw errors and these operations are usually successful. This produces a *corrected* BER value for the drive which is often in the range of 10^{-14} to 10^{-15}. The remaining bit recovery errors are errors that the drive cannot correct and are considered *unrecoverable* read errors. These are only discovered when read operations fail and can present data administrators with maximum career-surprise elements if compensating considerations have not been provided.

Obviously, the more bits a drive stores, the greater the probability an unrecoverable error can occur on the drive. Exacerbating this concern is that an inexpensive desktop-class hard disk can not only have eight times the capacity of an enterprise-class disk, but also usually has a BER rating of 10^{-14}, while an enterprise disk has a BER rating of 10^{-15} (the enterprise-class BER is ten times better). Hence, although it is important to test newly manufactured drives for unrecoverable errors and to remove bad detected media surface areas from use before they are sold to customers, it is even more important to do this when using desktop-class disk drives.

To accomplish this, a drive must write data to a data sector and then attempt to read it back to verify what it wrote is what it read back. When an error is detected, the sector is flagged as defective and not used again in lieu of an assigned alternate sector. Subsequent reads/writes to the defective sector address are then redirected to the alternate sector and the address is saved in a list of defective sectors known as the defect P-List (primary list). However, because desktop drives are now approaching half a terabyte, this read/write testing requires substantial time for an industry that survives on frugality and in which defects can appear after manufacturing testing. These subsequent defects are saved in another defect list and are also assigned alternate sectors.

Moreover, a disk drive's media recording surface might be vulnerable to certain data patterns but not other ones due to read/write head automatic gain control circuitry effects that vary with different data patterns. So one good question is, *what data pattern(s) should be written?*

The common denominator in any media defect testing is time, which is a direct function of a disk's capacity, sustained read/write transfer rate and how thoroughly the capacity is tested using different data patterns. For example, it can take 10 hours to test a single 80 GByte drive thoroughly.

Disk drive vendors spend relatively less time testing lower-margin, inexpensive desktop drives than enterprise drives because

- enterprise class drives have smaller capacities than desktop drives;
- margins are lower on desktop systems;
- time is money.

7.1.5 Mean time before failure (MTBF)

A useful discussion of MTBF is at: http://en.wikipedia.org/wiki/MTBF

In short, a hard disk's MTBF provides a manufacturer's informed *notions* of how long a drive design might be expected to operate within its recommended operating environment.

Assume a new disk model has a 400 000 hour MTBF. One way to interpret this MTBF value is that if you had 1000 disks of this model and subjected them to their designed duty cycle, 500 (half) of these disks would be operational after 400 000 hours and 500 would have failed. Note that there is no indication of when the failing disks became inoperative or what caused their failure.

Noting that 400 000 hours exceeds a 45 year interval, it is doubtful that any present drive will be in use in 45 years. The possibility is therefore nonexistent that a vendor has tested a disk drive model for 45 years and can vouch with complete authority that it will endure that length of time. Hence, it is important to note that a MTBF value is a statistical assertion that is necessarily based on a relatively brief testing period, vendor experience, physical handling and operational expectations, and individual component life cycle projections. There are a few more points to consider with the MTBF metric.

Vendors assume enterprise and desktop operational environments are different. Enterprise drives are assumed to operate continuously – 24×365 – while desktop drives are assumed to operate 8×5. Moreover, enterprise drives utilise more rugged mechanical components and are constructed in ways that allow them to provide higher relative performance in ways that generate less wear and are more forgiving.

For example, enterprise drives usually rotate faster than their desktop counterparts. Consequently, 3.5 inch form-factor enterprise disks often use internal storage platters that are smaller than maximally possible in order to reduce air friction heating. This heating increases exponentially with higher RPM rates due to rotational kinetics. The platter diameter reduction reduces the available data encoding surface area, commensurately reducing storage capacity and consequently associated seek distances. Moreover, due to higher rotation rates and to achieve higher read/write margin tolerance, areal density is also reduced on enterprise class drives. This further reduces storage capacity.

Finally, a major source of drive failure stems from actuator wear. Vendors usually have a fairly good understanding of how fast actuators can wear under assumed operational loads. But at least one major vendor declines to reveal their data. Yet, it is safe to assume that desktop drives that operate in continuous 24×365 heavy seek environments will fail much sooner than their enterprise class counterparts.

Indeed, one Symantec Senior Principal Software Engineer has a testing rack of 160 blade systems, each with an embedded 2.5 inch IDE drive. The rack has been continuously housed in an air conditioned lab. Though these drives have been powered on (i.e. have been continuously spinning) for over three years, they have been subjected to very light I/O activity. Nonetheless, it is not unusual for three to four drives out of the 160 to fail in a week. In contrast, the engineer has had approximately a total of three or four SCSI drives fail under similar circumstances in the same three years.

7.1.6 Failure rate breakdown

In his 9 March 2005 *Storage Network Conference* panel discussion titled 'New Directions in Storage Components and Systems', Gordon F. Hughes presented a recent ACM paper he co-authored with Joseph F. Murray from the University of California San Diego. This paper is highly recommended (see Bibliography at the end of this chapter) and provides a discussion of hard disk failure analysis.

The paper provides many important observations, and we mention just two here. First, desktop drives are not as rigorously tested in manufacturing as enterprise drives because of their significantly larger capacities, slower sustained transfer rates and commensurate testing time required. It simply costs too much to perform a comprehensive surface analysis of each desktop drive and it is not a good bet to expect it.

Second, while enterprise drives are returned to manufacturers, most desktop drives are simply discarded and replaced because of their low cost. Hence enterprise drive failures are more completely understood than desktop drive failures.

From the paper, Table 7.1 is the breakdown of drive failures based on an analysis of 4000 enterprise drive failures one vendor experienced:

Table 7.1 Breakdown of drive failures

Failure mode	Description	Frequency (%)	Stress Condition
Head-Disk interference	Head crash	15.5	Operating
No problem found	Returned drive tests OK	15.0	–
Recording heads	Complex nanotech devices	14.5	Operating
Post manufacture	Drive handling damage	10.1	Non-Op
Circuit board 'PCB'	Many IC components	8.5	Operating
Head or disk corrosion	Causes disk HDI or defects	7.7	Non-Op
Head assembly 'E-Block'	Wires, preamp, coil	6.8	Operating
Head Disk Assembly	mechanics, electrical, voice coil	3.9	Operating
Disk defects	Causes HDI or read errors	2.6	Operating
Drive firmware	Internal operating system	1.9	Operating
Head-Disk stiction	Disk won't spin up in drive	1.3	Non-Op
Spindle bearing	Disk spin bearing	1.1	Operating
Contamination inside drive	Foreign gases or chemicals	0.7	Op/Non-Op

7.1.7 No free lunch

In the final analysis, the maxim *you get what you pay for* holds true
when it comes to present hard disk reliability. Lower cost disks
should be expected to have lower reliability and they usually do
for a spectrum of understandable economic reasons.

Therefore, in summary:

• desktop drives are usually lower cost drives;
• lower cost drives have lower reliability;

- lower cost drives, hence lower reliability drives, are usually Serial ATA drives.

However, even today's lower reliability disks have reliabilities that approach yesteryear's higher reliability drive levels. And, the real question is:

> Can these lower reliability drives be applied in ways that exploit their larger storage capacities while economically compensating for their individual lower reliability in ways that render their collective reliability sufficient?

Here, the answer is a distinct *yes*.

7.2 RAID review

Although the acronym RAID now stands for Redundant Array of *Independent* Disks, recall that the original 1987 acronym RAID stood for Redundant Array of *Inexpensive* Disks. Back then, inexpensive disks were fairly unreliable when compared to mainframe storage devices and the original meaning endured until RAID technology enabled small, 5.25 inch, inexpensive disks to essentially obliterate much larger and much more expensive enterprise-class storage counterparts. So, historically, RAID technology has played an important role in the quest for reliable storage subsystems that enable enterprises to accommodate exploding data growth.

Today, we are reliving a variant of the earlier battle between large expensive disks and RAID configurations using small inexpensive disks. Only this time, the battle engages disk offerings that have the same physical size but different reliabilities. Indeed, when enterprises now build large storage arrays to store critical data, they tend to want to use inexpensive desktop-class drives for budgetary reasons. But, as already discussed, the unrecoverable read error rate desktop class drives experience, combined with their enormous capacities, render them generally unsuitable for this purpose without compensating mechanisms.

Hence, RAID comes to the rescue once again.

7.2.1 RAID 5 review

RAID 5 partitions the data storage capacity of several drives into equal-sized *stripes*. A stripe comprises an equal-size fraction of storage capacity, a *strip*, from each participating drive. Within a stripe, one strip is reserved for storing a byte-wise parity calculation that is calculated by byte-wise XORing the data values residing in the remaining strips housing data. Should one of the stripe's strips become unavailable due any reason ranging from complete drive failure to media defects, the unavailable data can be reconstructed from the usable strips which often includes the parity strip for the stripe.

When a drive completely fails, it is necessary to replace the drive (perhaps with a hot-spare drive) and reconstruct the lost drive's contents. This reconstruction is called a *rebuild*. To prevent data loss, rebuilds must complete before encountering another drive failure or unrecoverable read error. Unfortunately, elementary mathematics suggests a rather unpleasant reality.

Keep in mind that a rebuild requires reading all data on all disks to perform the necessary recovery operation. Moreover, a RAID 5 array typically uses disks having the same capacity, and usually, the same host controller transfer performance. Since the disk reads can occur simultaneously, it is only necessary to determine how long it takes to read a single disk.

Suppose an array hard disk has a 400 GByte (400 000 Mbytes) capacity and a 40 MBytes/sec sustained host transfer rate. It follows that the fastest possible elapsed rebuild time is roughly:

$$(400\,000/40)\,\text{seconds} = 10\,000\,\text{seconds} = 2.78\,\text{hours}$$

For small arrays the array rebuild time is likely to be constrained by the sustained media transfer rate of the slowest disk. However, with large arrays (40+ disks) a RAID processor must transfer and XOR the data from all source disks. If the RAID array has, say, 40 400 GByte data disks, then this implies a total transfer of 16 TBytes (16 000 GBytes).

Now, today's XOR engines can often consume at most 500 Mbytes/sec of source data to generate parity information. So, the maths here is fairly easy.

$$400\,\text{GBytes} \times 40 = 16\,000\,000\,\text{MBytes}$$

$$16\,000\,000\,\text{MBytes}/(500\,\text{MBytes/sec}/(60\,\text{sec/min})/(60\,\text{min/hr})$$

$$= 8.89\,hours$$

So, with a 40-drive array of 400 GBytes source data drives, the RAID controller must transfer and XOR 40×400GBytes or

16 TBytes of data at, say, 500 MBytes/sec, which requires about nine hours. Finally, note that any RAID rebuild time may vary, probably for the worse, depending on the characteristics of the RAID subsystem and the impact of additional processing that the RAID subsystem performs. Remembering that disk failures can occur in clusters for disks in an array, an actual rebuild time might well be considered as a long, categorically unacceptable, time to expose enterprise data to a concurrent storage failure during a rebuild recovery operation.

In other words, because of the enormous storage capacity of today's SATA hard disks, RAID 5 is quite possibly *too slender of a reed* for enterprises to rely on for large SATA drive arrays requiring enterprise-class reliability. Hence the search for RAID 5 alternatives and the emergence of RAID 6.

7.2.2 RAID 6 overview

RAID 6 technology is referred to using a variety of names, including Dual Parity (DP) and RAID 5+DP. Although RAID 6 technical particulars are beyond the scope of this book, trust that RAID 6 is a logical extension of RAID 5 technology.

RAID 6 uses the same concept of RAID 5 stripes and strips. While a RAID 5 array can experience one strip failure without losing data, usually a second concurrent strip failure will *cause data loss*. With RAID 6, an array can experience *multiple* concurrent data failure occurrences without losing data. Indeed, RAID 6 enables data administrators to select how many concurrent data failures a given RAID 6 array can experience before losing data. This is accomplished by providing multiple sets of independent stripe recovery information within the array.

For example, if an administrator installs an array with 40 equal-capacity drives and desires to withstand *two* concurrent data failures, there will be two sets of independent recovery information within the array, leaving a 38-drive storage capacity for data. If an administrator installs an array with 40 drives and desires to withstand *three* concurrent data failures, there will be three sets of independent recovery information within the array, leaving a 37-drive storage capacity for data.

In RAID 5, the parity information consumes the capacity of one drive in the array. Similarly, each set of independent recovery information within a RAID 6 array consumes the storage capacity of one drive array regardless of the data and recovery information placement strategy within the RAID stripe.

So, RAID 6 offers enterprises an opportunity to exploit SATA desktop drive capacities within arrays that must exhibit enterprise data availability, despite SATA drives inherently having less reliability. In fact, consider that a 60-drive RAID 5 array might suffer a data loss every six months but that using two independent sets of recovery information with RAID 6, the same 60 drives would nominally experience a data failure every 100 years. Not bad – trading some additional processing time and the capacity of one drive for a 200X improvement in data reliability.

The final topic in this chapter reveals how to combine the benefits of tiered storage and object-based data classification

7.3   Tape-based solutions

Any data management discussion is incomplete without discussing tape storage's new role in the lifecycle process. In this process, tape storage has generally been used as an offline storage medium or *secondary storage*. However, since the introduction of *Virtual Tape Libraries*, this process has become somewhat modified. Concurrent with the introduction of newer disk drive technology, the storage industry has also entered a new era with tape. The resulting *Virtual Tape Library* is a hybrid technology that is garnering impressive support because of its ability to deliver advantages within certain applications.

Historically, tape has only been considered as a backup or archive medium. Nonetheless, it is still considered a primary vehicle for managing data throughout its lifecycle. As an archive medium, it has supported *Hierarchal Storage Management* (HSM) applications that automate data objects migration and retrieval processes. Its use as a HSM migration tool naturally encouraged strategies for segregating data objects by usage mode and access frequency. And, in most cases, data object copies went to 'secondary' or tape locations. Coupled with an advanced backup technology, the storage industry was able to provide enhanced integrated data management suites.

However, augmenting tape storage resources with lower-cost disk resources assists in reducing tape operation costs as well as tape use and its associated libraries. Utilising and combining lower-cost disks with lower-cost tape devices therefore enabled enterprises to capitalise on tape's greatest strengths. This reduced operational costs by maximising effective tape utilisation, reducing the number of tape appliances and cassettes, and adding additional value by providing enabling lifecycle processes.

7.3.1 Virtual tape library primer

Virtual tape library technology allows faster and more reliable backups using disk-based storage. This technology is not new, but enhancements regularly continue to appear. In the past, so-called *Virtual Tape Systems* were library products that used disk buffers to contain data until it filled a tape. The tape was then written to full-capacity.

A virtual tape library emulates tape medium and tape libraries. Since virtual tape libraries use disk storage, backups are generally faster than having to load and position tape medium. However, newer and contemporary tape devices are now available such as Linear Tape Open (LTO). Backed by a triad of companies that include IBM, Hewlett-Packard and Seagate Technology, LTO products are now available that provide greater tape speeds at a positive nonlinear rate, pushing new disk technology capabilities.

On the opposite side of the spectrum, the instant rewind and instant seek times do not compare with that of disk devices. Moreover, mechanically intensive operations such as loading, unloading and ejection are virtually eliminated by using virtual tape libraries, this significantly reduces operational inconvenience and frustration.

Virtual tape library technology works by presenting disk storage to network systems as a tape library. Depending on the technology or hardware, this presentation could comprise different vendor devices. Disk devices are presented as a tape medium and can be made available to different network hosts. Supporting multiple simultaneous tape backup sessions enables enterprises to provide 24×7 backup policies (at the expense of potentially longer individual backup times)

The virtualisation strategy can be expanded by having the virtual tape library (VTL) appear as different units. An example VTL is shown in Figure 7.1.

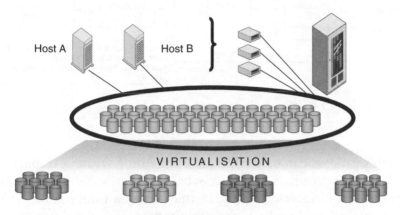

Figure 7.1 Virtual tape library

The complex that Figure 7.1 depicts could theoretically be used by sharing the virtual tape devices with multiple hosts. For example, a VTL created by a software layer could provide shared access to virtual tape resources by using the shared disk and the shared disks on a Storage Area Network could be accessed and used by multiple hosts on the storage area network. This essentially creates a virtual storage pool paradigm with various storage elements which can be drawn dynamically. The storage pool can be used by backup software to effect lifecycle policies. Depending on requirements, backups can be initially sent to virtual tapes utilising different disk tiers and then duplicated to one or more physical tapes for onsite or offsite retention supporting disaster recovery or deep data archive purposes.

As mentioned above, even newer roles are appearing for tape. Hence, it is likely that tape will remain a viable lifecycle medium for the next few years. Moreover, we will either see more use of VTL's or advanced volume management virtualisation capabilities within backup software to allow for similar offerings.

Figure 7.2 depicts a conceptual view of a process utilising a virtual tape library or a tiered infrastructure.

In Figure 7.2, data movement between the disk-based virtual tapes and physical tape media is accomplished as a secondary

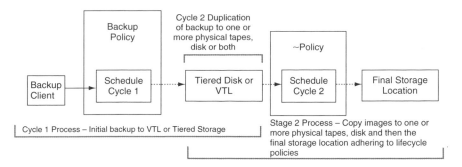

Figure 7.2 Example lifecycle data backup process

task; this process can be completed at any time without impacting application performance. Moreover, it allows more granular approaches to lifecycle polices. In that sense, a policy could exist that contains information describing how the data would be moved, copied, duplicated or archived and retention times associated with the data. In the end, by utilising such a disk and tape hybrid resource, costs could be curbed in either disk or tape direction.

However, deploying VTL's requires a thorough requirement assessment. The backup applications should be clearly understood and analysed to give more realistic hardware and software requirements for meeting business and company level objectives.

Although initial lifecycle management involves most parts of the storage (and even server and application) technology stack, there are particular areas in which companies should focus their initial efforts to enable their environment to introduce lifecycle processes. Attempting to produce a comprehensive framework will be difficult when attempted in an single implementation. In many cases, initial adoption could likely proceed on a discrete basis within a company – focusing on a particular application, compliance requirement or business process/workflow, and subsequently being widened to incorporate additional applications or processes.

Accordingly, enterprises should focus on initially supporting individual lucrative applications in response to their internal demands, and then generalising the expertise acquired for use with the wider enterprise. This is where virtualisation software or hardware in the form of VTL's could prove advantageous.

7.4   Recoverability of data: you get what you pay for

The preceding sections described the differences between Serial ATA and ATA disk devices and their practicality and feasibility in an enterprise environment. More importantly, applying certain RAID technology can leverage the lower cost per GByte drives in our tiered storage model that is part of a bigger lifecycle picture. Going further, clearly understanding the capabilities of these different storage mediums, provides a foundation to build a lifecycle strategy that assists in being more competitive in the market place. That said, our discussion also examined the new *Virtual Tape Libraries* tape uses within lifecycle policies. Again, data has different value over time and having the ability to transition this data over storage tiers – online, near-line or offline – is critical.

Now, consider applying software technology concepts and objected oriented processes to disk and tape hardware that allows data to be effectively located, organised, moved, transitioned, migrated, archived and recalled. The classification process and concepts discussed previously assist in determining where the data should reside based upon the data criticality. Using the different data classes, we can theoretically store the data on different storage tiers. However, the data criticality differs from class to class.

The identification and classification process is extremely beneficial in determining where the data should reside now or in the future. The preceding discussion pertaining to classification mentioned that the data should be visible to the storage and data management tools available. By carefully organising data objects, a company can choose the best strategy to suit its needs.

For example, Figure 7.3 depicts a view of a *Data Source Object* (DSO). The DSO could be organised at a granular level, permitting it to be moved to various storage tiers based upon cost/performance. Migration from a particular storage tier to another tier could be automatic and transparent, according to usage or user-defined based policies.

Data lifecycle management requires storage administrators to plan and implement policies and procedures with the business that organises data, controls access to it as appropriate, retains or removes it as appropriate, and allows content searches on all types of data. Ideally, after classifying data (based on criteria such as its criticality or age) administrators will also be able to

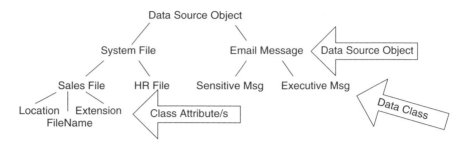

Figure 7.3 Data Source Objects

store different classes of data on different classes of storage tiers. This management should provide means to differentiate activities by required media performance, availability and other necessary criteria. Because required data format changes are always possible, the means need to accommodate these changes.

Advanced file system, filter drivers or volume virtualisation software can aid moving DSO's to a particular data tier and classifying it according to the predefined policies that a company selects. A company should be able to construct polices and DSO jobs according to data classifications. For example, 'HR data' can reside on a specific, designated tier. Finally, note that different lifecycle policies and jobs may act on many storage volumes at a given scheduled time or manually.

To enable data object movement to intended storage locations, storage administrators define policies that allow lifecycle policies to select files based on the policies and transition them to their pre-determined location. Administrators could theoretically send specific types of data (such as images) to specific media, send older data to specific media, or exclude specific data from movement. Figure 7.4 gives us a high-level view with DSO allocation.

In this scenario, placement choice is essential. Figures 7.5 and 7.6 depict example movements of files across the different tiers, including a tape tier. Here, lifecycle polices could move data across different hard-disk storage tiers and then finally to a tape tier for deep archive. Each policy would have different retention times based upon the data business criticality. As this section's heading mentioned, you get what you pay for – especially during recall operations.

How the various capabilities of the technology in the enterprise are used and implemented is very dependent on the specific need to recover, recall, migrate, move, transition or archive, balanced

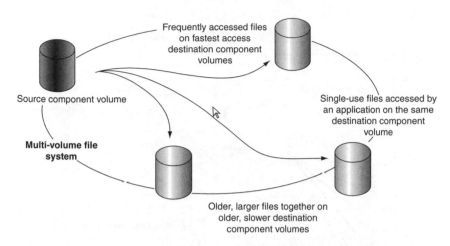

Figure 7.4 Data Source Objects over virtual storage volumes

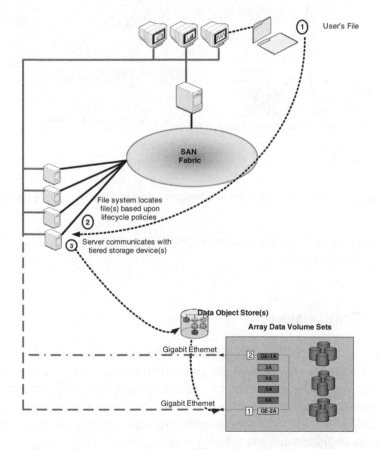

Figure 7.5 Data Storage Objects with storage tier

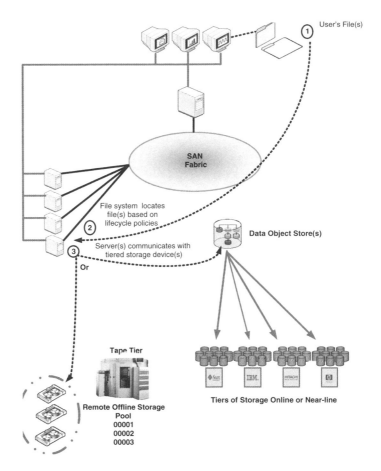

Figure 7.6 Data Storage Objects movements across tiers

against the cost of the various hardware, software and business service level agreements.

7.5 Conclusion

Utilising the various file systems, virtualisation or RAID techniques, can clearly benefit an overall lifecycle process. A storage administrator designing and implementing a lifecycle strategy can take several steps to enhance the contribution made by the various technologies and processes. Significant cost reductions are possible without indiscriminately placing data at risk.

The following questions are appropriate to ask when an enterprise considers that using a tiered storage should be part of its strategy:

- Does the storage strategy include short and long-term storage of data?
- Must the enterprise maintain data for extended periods of time?
- Does the data fluctuate and have different requirements?
- Is the primary storage capacity reaching its utilisation limit?

If the answers to these questions are *yes*, the enterprise should seriously consider a tiered storage model.

Bibliography

Gelsinger, P. (1999) http://www.intel.com/pressroom/archive/speeches/dpo22399.htm

Hughes, Gordon F., and Joseph F. Murray (2004) *Reliability and Security of RAID Storage Systems and D2D Archives Using SATA Disk Drives,* University of California, San Diego, *ACM Transactions on Storage,* **1**(1), 95–107.

Schwaderer, W. David (2003) *Serial SANs – The Convergence of Storage and Networking,* unpublished manuscript, excerpted by permission from the author, all rights reserved.

Schwaderer, W. David (2005) *Innovation Survival – Concept, Courage, and Change,* unpublished manuscript, excerpted by permission from the author, all rights reserved.

Toigo, J. (2000)*The Holy Grail of Data Storage Management: Modeling, Evaluating, Comparing, Implementing Storage Management Options and Solutions.* Upper Saddle River, New Jersey, Prentice-Hall.

VERITAS Software Corporation (2004) *Delivering Quality of Storage Service, Using a multi-dimensional storage hierarchy to optimise the performance, availability, and cost of storing digital data,* VAN article archives.

8

Continuous Data Protection (CDP)

8.1 Introduction

Continuous Data Protection (CDP) is an emerging technology that enables enterprises to increase their ability to provide sustained application and data availability through enhanced data recovery capabilities. Within the data recovery context, the Storage Networking Industry Association (SNIA) defines CDP as:

> *A data protection service that captures changes to data to a separate storage location. There are multiple methods for capturing the continuous changes involving different technologies that serve different needs. CDP-based solutions can provide fine granularities in time or versions of restorable objects ranging from crash-consistent images to logical objects such as files, mail boxes, messages, etc.*

The first thing to note within this definition is its classifying *files, mail boxes, messages* as *objects* that are available for recovery. A second thing to note is the complete absence of suggesting how data changes are captured, reflecting the multiple levels at which of CDP data can be captured: application (transaction) level, file level or volume level.

CDP is a front-end protection system that is always on, operating unobtrusively to enterprise applications. A typical CDP system is based on a disk storage infrastructure to log the continuous

Data Lifecycles: Managing Data for Strategic Advantage   Roger Reid, Gareth Fraser-King and W. David Schwaderer   © 2007 VERITAS Software Corporation. All rights reserved.

data changes as well as provide a time indexed view into historic points in time. As such CDP systems may require additional processing resources depending on the CDP approach. In return, CDP can provide enterprise IT organizations with seamless, near-instantaneous recoveries from logical and physical data corruption events stemming from many sources, including operator errors. By 'near-instantaneous' we imply recovery in seconds or minutes rather than the hours that a traditional application and data restoration operation may entail.

This near-instantaneous recovery is achievable because CDP effectively produces more granular recovery points that collectively result in continuous data instances that are 'instantaneously' available on the CDP disk storage, sometimes referred to as 'CDP Sideview'. A CDP recovery operation appears as a restoration to *Any Point in Time* (APIT) – a selectable time point that varies by application and operating system. APIT recovery capability therefore enables enterprises to capture continuous data integrity recovery information and restore application services to a point close in time with a data corruption event.

Although RAID technology and remote mirroring can provide enterprises with *physical* data protection by virtue of device redundancy, CDP correspondingly provides for both *physical* and *logical* data protection. Because CDP continuously operates under the covers, no scheduling is required and inconvenient backup window considerations are avoided. Since its focus is on near-instantaneous recovery, it follows that CDP expands rather than supersedes the data protection spectrum that legacy data protection and disaster recovery services already provide to enterprise users.

To illustrate this, Figure 8.1 shows where CDP fits from both a RPO and RTO perspective, relative to other current protection/recovery technologies like snapshots, disk backup and tape backup. (The Recovery Point Objective, or *RPO*, is a measure of how much data loss due to a node failure is acceptable to the business. A large RPO means that the business can tolerate a great deal of lost data, and a small RPO means that businesses can't tolerate a great deal of lost data. The Recovery Time Objective, or *RTO*, is a measure of the users' tolerance to down time. A large RTO means that users can tolerate extensive down time, and a small RTO means that users can't tolerate extensive down time.) In addition to providing a more granular RPO and RTO and thereby reducing

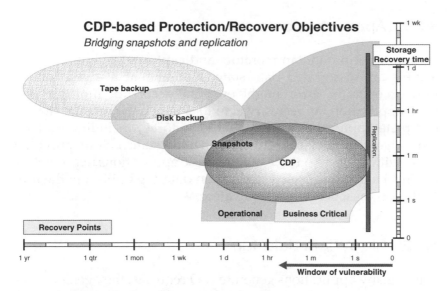

Figure 8.1 Integrating CDP, ILM and tiered storage

the window of vulnerability of data-loss, CDP by virtue of its similarity of data capture method, bridges the gap with replication

A CDP solution can provide not only historic points in time but also, in most cases, has 'current' images or replica, which is the most up to date storage instance. Several in-market CDP solutions today are derivatives of replication products and often provide dual purpose capabilities.

In the next couple of sections we break down CDP into its components and look at CDP-based protection and recovery data flow.

8.2 CDP data-taps

CDP operations typically begin by intercepting data change activity at one of three *points* that each comprise an I/O subsystem stack interface. The software that accomplishes this is called a *data-tap*. In decreasing levels of application awareness, the three tap points are:

1. Application Data-tap;
2. File System Data-tap;
3. Volume Data-tap.

8.2.1 Application data-tap

An application data-tap monitors and captures application trans-actions and associated data state. Noting that a given transac-tion or operator error might eventually corrupt multiple files, CDP application tap recovery activities necessarily ensure that the complete application working set is restored, including buffered data changes, to a collectively coherent state before providing users with application access. Enterprise applications (e.g. database engines) may have such a application data-tap built-in in the form of their transaction log or equivalent.

8.2.2 File system data-tap

When many applications generate I/O requests, they generate and present a request (file open, file close, file read, file write, set file position for next operation, etc.) to the operating system using stan-dardised operating system requests. The operating system exam-ines all requests and determines when a given request should be passed onto the file system. The file system data-tap point there-fore enables CDP file system tap software to monitor and capture all file system activity.

As an example, operating systems such as Microsoft's NT® and XP® operating systems use an Installable File System (IFS) archi-tecture that enables a file system to provide services through a well-defined interface. Here, the IFS interface provides an installed file system with structured requests and the file system Microsoft provides is the NT File System (NTFS).

8.2.3 Volume data-tap

At a data volume level, CDP volume data-tap software monitors and captures all data block changes at the volume block level. Since application consistency isn't guaranteed through any application constructs, additional provisions are required to assure write order fidelity. Additionally, if an enterprise application spans multiple clients/hosts and storage groups, the volume level data-tap data flow needs to be captured in a so-called CDP consistency group. The CDP consistency group guarantees write order fidelity either at the block level transaction or a very narrow consistency interval.

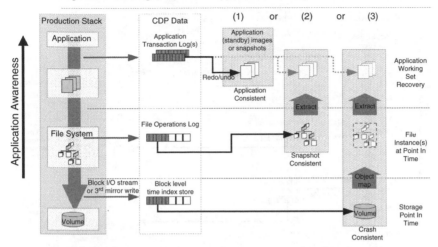

Figure 8.2 Continuous protection and recovery flow

As Figure 8.2 depicts, the three different tap points enable CDP systems with different application awareness levels. Moreover, the figure illustrates the relationships that need to be established with higher levels of I/O subsystem stack in order to provide a recoverable application working set. That is, if the objective is to recover a consistent application working set, then with application-based CDP, you'll get that almost automatically, albeit at the possible cost of another complete application infrastructure. For file-based and volume-based CDP, additional recovery steps are required, equivalent to crash recovery or file system journal-based recovery, to get to a consistent storage image, file system state and application working set.

Given the different levels of CDP data-tap, in reality this will result in different types of data being captured continuously. As a result, the corresponding CDP store typically is optimised for the data type that is being captured and for the CDP APIT instance that it needs to export/share: an application working (file/data/database) set, file instances or volumes/LUNs.

In Figure 8.3, the different types of CDP stores are a consequence of the

- different data streams originating from the different CDP data-tap types;

CDP Store Types

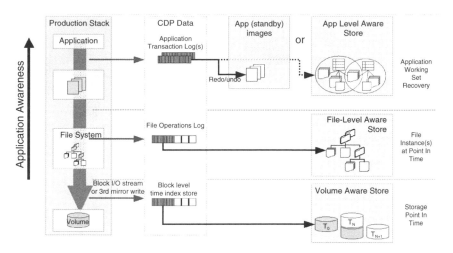

Figure 8.3 CDP store types

Table 8.1 CDP data-tap advantages and disadvantages

Data-tap Level	Advantages	Disadvantages
Application	Consistent granular transaction redo and undo Part of application Data Recovery solution Metadata knowledge and versions	2nd application licence required Different solution per application
File System	Granular/selective restore Metadata knowledge and versions Ability to provide index and search as well as cataloguing May include in-memory state	No guaranteed application consistency Fragile reverse engineering to make application-aware Requires host presence
Volume	Can get APIT image Production host transparency Broad application support Drive replication convergence	Requires crash recovery Difficult to index and search by time No implicit object level catalogue Reverse engineering for object recovery

- optimisation of the actual objects (transactions, file/message objects, file operations or volume blocks) in the CDP store;
- ability to create efficient recovery points in time and/or versions of the stored objects or images.

Table 8.1 provides the advantages and disadvantages of each type CDP tap point.

8.3   CDP operations

CDP logging activities necessarily use disk storage resources to perform all logging activities. CDP 'disk buckets' are typically near-line storage resources within a tiered storage environment. By avoiding tape technology, data administrators can provide faster recovery operations and higher reliability while avoiding traditional tape media problems. Within many multi-host environments, storage administrators can also mount CDP activity logs as read-only resources and rehearse CDP based recovery and test operations (e.g. disaster recovery (DR) testing) using mirror copies without interfering with production activities.

Figure 8.4 depicts a traditional enterprise storage network with a shared SAN disk pool (FC or IP) that provides capacity for SAN and LAN volume backup and recovery, extended with a typical CDP infrastructure. This storage infrastructure complex also provides backup and recovery capability, with for example, VERITAS NetBackup™.

A CDP solution consists of three components:

1. A CDP software *data-tap* at a production host(s).
2. A *CDP Store* that provides disk storage for CDP logging activity, typically using existing shared SAN storage resources. Architecturally the CDP server as shown, may be an out-of-band appliance or in-band appliance/gateway.
3. CDP *consumer applications* that access CDP logs/instances in addition to normal CDP facilities. Typically the CDP consumers connect to the 'CDP Store' through the CDP server which exports/shares virtual time instances (APIT) of its corresponding data type (application instance, file-system or volume/LUN).

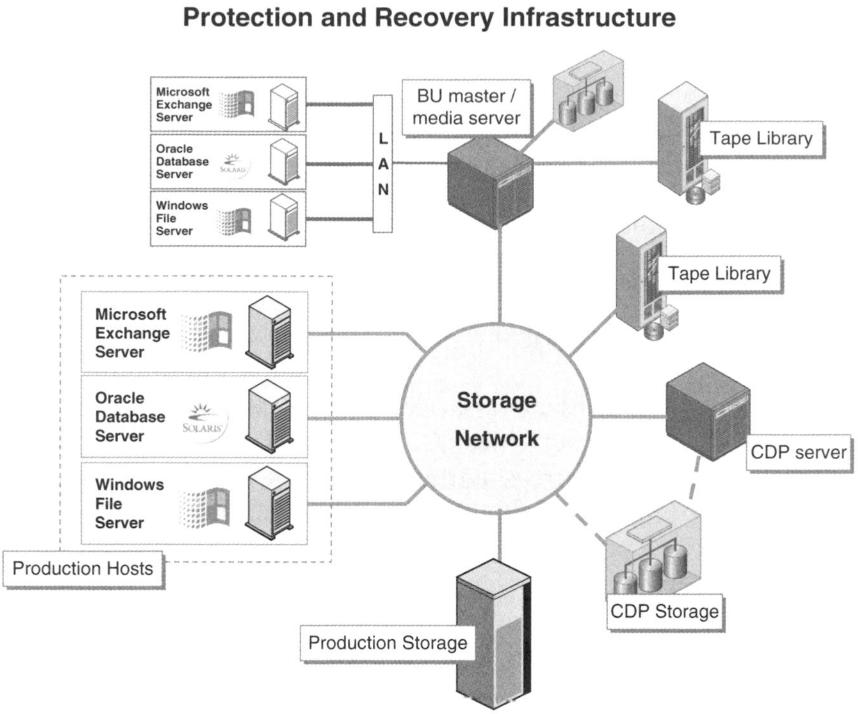

Protection and Recovery Infrastructure

Figure 8.4 Traditional enterprise SAN configuration with CDP capability

Introducing a CDP capability therefore results in the configuration depicted in Figure 8.4.

When reviewing data protection and recovery using the CDP infrastructure, we'll start with the data-tap. As Figure 8.5 depicts, a CDP software tap captures the I/O stream at one of three potential points within the production host and in parallel to production I/O. The data-tap writes captured change information to the CDP Store in a continuous, append-only mode. The accumulated information and an original base-line data snapshot provide the I/O state for APIT recovery as long as a continuous I/O stream is retained.

When corruption is detected, *near-instantaneous recovery* is possible (i.e. without data restoration or data movement) by failing-over or remounting an exported/shared CDP Store APIT image as a LUN, volume, file-share, mail-database, etc. For example, data administrators can restart or recover a current

CDP Data-tap

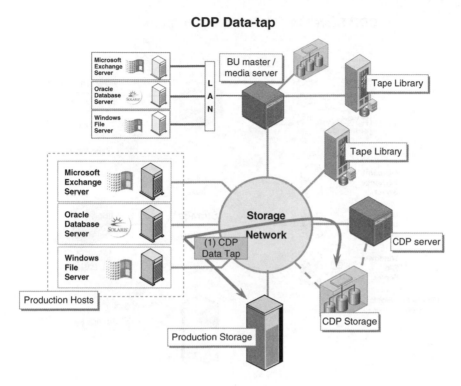

Figure 8.5 Tap data change capture

Oracle database instance with a state prior to an accidental table purge by using a CDP APIT and performing an Oracle crash recovery from that image.

8.3.1 CDP store

The CDP Store has the following attributes (Figure 8.6):

- Time-indexed store; APIT image or individual data instances;
- Store optimised by data-tap type:

  - exports LUNs for block-level CDP;
  - exports File System share for file level CDP;
  - custom application object store.

CDP Store

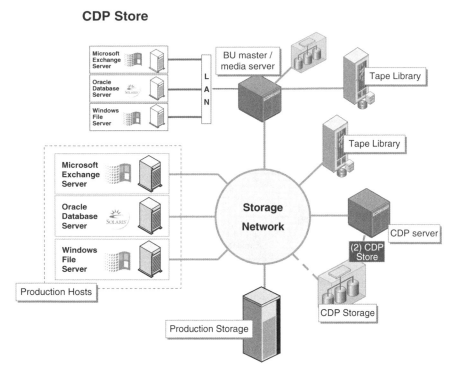

Figure 8.6 CDP Store

The CDP Store (Figure 8.7)

• has the ability to resume operations immediately;
• typically requires R/W access to a temporary volume;
• has background production storage re-sync (e.g. restart Oracle before table purge).

In addition to rapid CDP host application recovery, the CDP Store's independence enables other backup/recovery applications to provide additional recovery services (Figure 8.8) Here, any data management application can *instantly access* APIT storage images or objects and restore a subset (a file; directory, mailbox, mail-message, table, etc.) to production storage and then continue or restart the current production application. This access and restore activity can occur off-host (in this example, not on the NBU media server) or on the production host (e.g. simple file copy). This

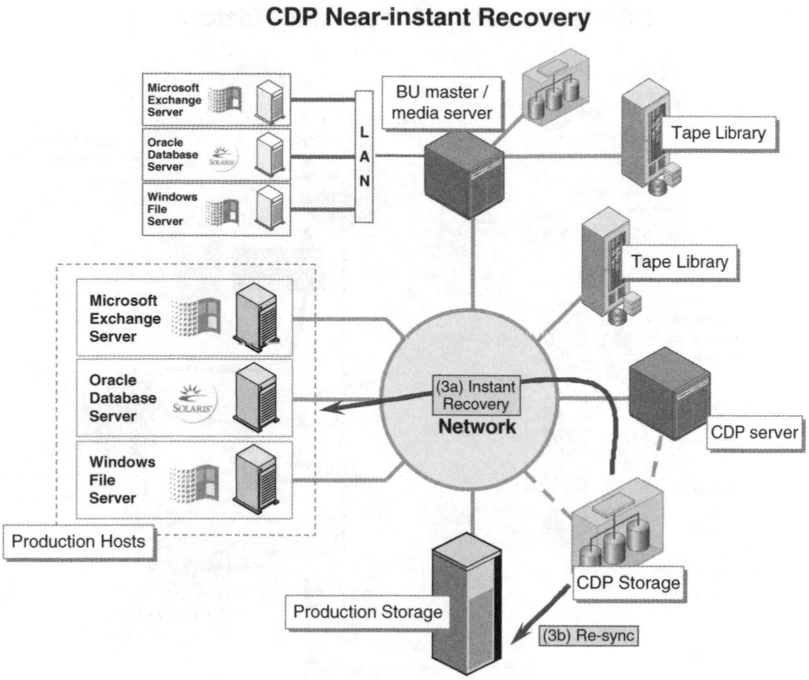

CDP Near-instant Recovery

Figure 8.7 Near-instant recovery

enables other applications to access, say, historic APIT Exchange database images and restore a complete/partial mailbox or a production Exchange database that was previously corrupted or deleted.

The CDP Store

- has the ability to instantly provide (read-only) access to image or object(s);

- requires a complete/partial restore step prior to continue/resume operation (e.g. restore mailbox from an hour ago).

In addition to providing APIT data protection/recovery, Figure 8.9 depicts how additional consumer applications can examine a CDP Store for second stage (traditional) backup, archival, data compression, SIS processing, data intelligence, second instance database processing for test and development, content indexing, object/version cataloguing, remote replication,

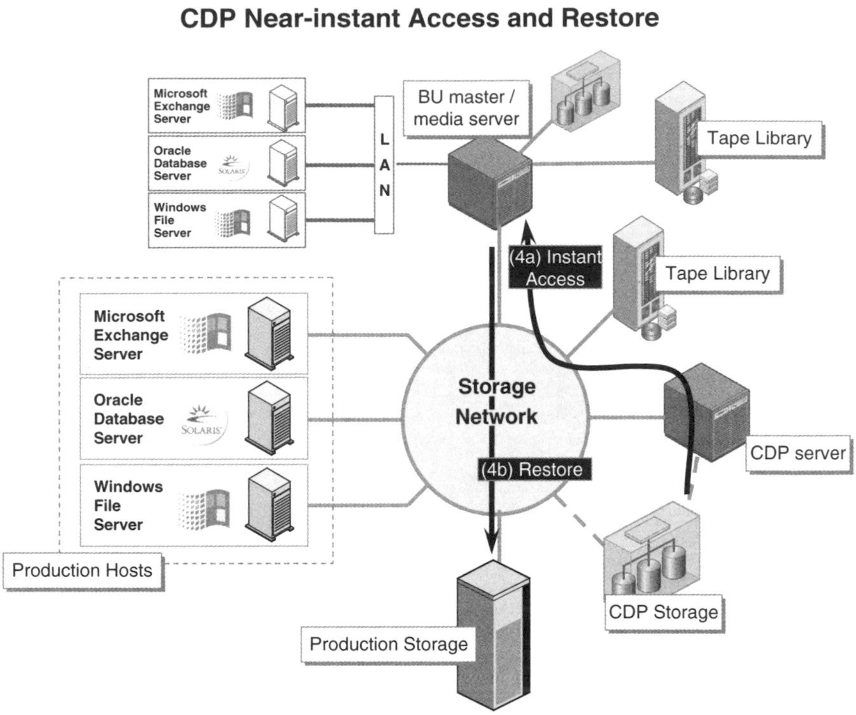

CDP Near-instant Access and Restore

Figure 8.8 Near-instantaneous access to APIT data by additional consumer applications

etc. Such CDP side-band applications often prove invaluable, especially when done without impacting the production host.

- tertiary media backup (tape or otherwise);
- archival, retention management;
- content Indexing, Search, Catalogue;
- data compression, SIS and remote replication;
- decision Support, Test Host.

8.3.2 CDP stakeholders

From the previous discussion, it is clear that a major CDP benefit is its ability to support a variety of off-host consumer applications. And, beyond CDP's substantial protection and recovery capability, this consequence may provide the biggest justification for adopting

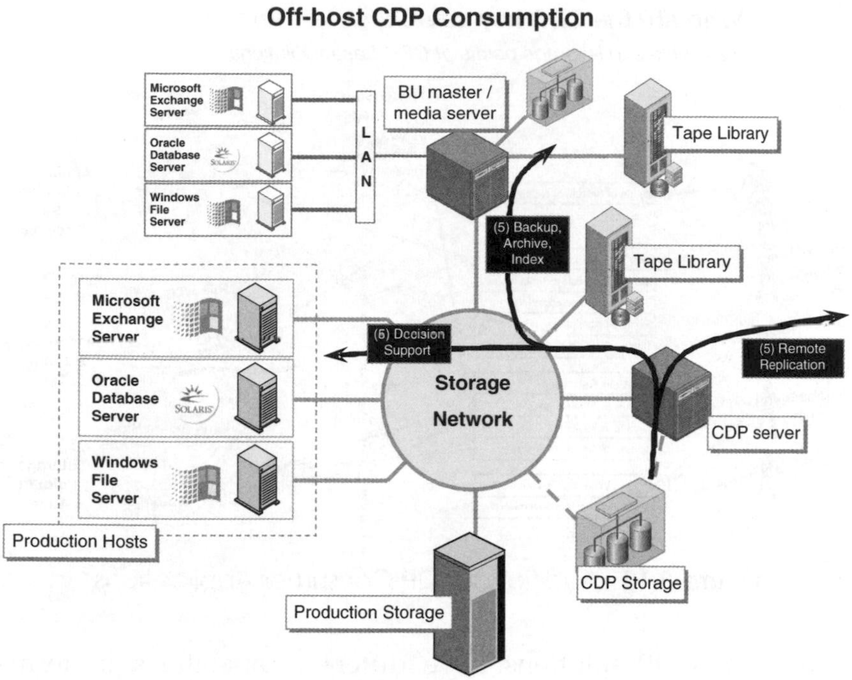

Off-host CDP Consumption

Figure 8.9 — diagram with the following labels: Microsoft Exchange Server, Oracle Database Server, Windows File Server, LAN, BU master / media server, Tape Library, (5) Backup, Archive, Index, Tape Library, Microsoft Exchange Server, Oracle Database Server, Windows File Server, Production Hosts, (6) Decision Support, Storage Network, (5) Remote Replication, CDP server, CDP Storage, Production Storage.

Figure 8.9 Additional CDP consumer applications

CDP technology. It follows that CDP implementations usually have numerous interested stakeholders. These include

- end users;
- archivists;
- Exchange/NAS administrators;
- DB administrators;
- Storage administrators.

Figure 8.10 depicts the complex relationships and dependencies various stakeholders create as they view their storage requirements and functional expectations through different lenses.

Without a doubt, Figure 8.10 underscores the environmental complexity different CDP solutions must address. Clearly

- Different stakeholders have different expectations based on their specific recovery requirements.
- Different applications have different (technology) requirements, in terms of brick level or storage level recovery.

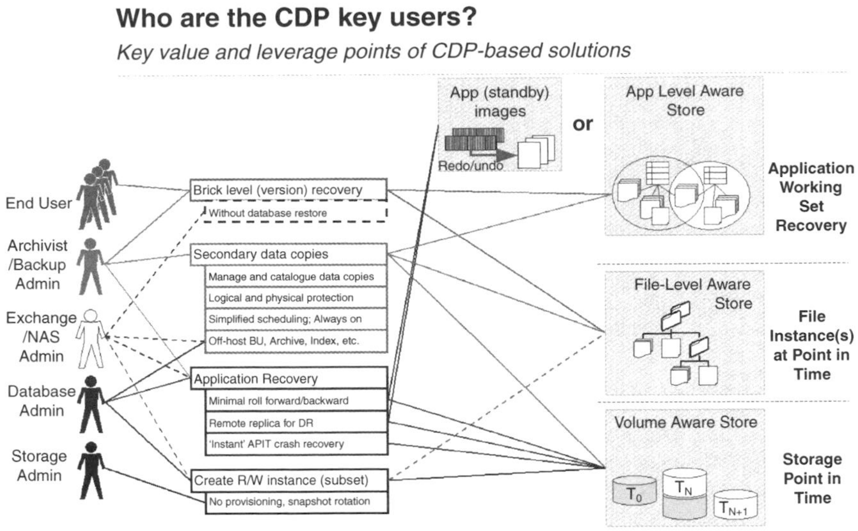

Who are the CDP key users?

Key value and leverage points of CDP-based solutions

Figure 8.10 Additional CDP Consumer Applications

- Different CDP solutions have different capabilities in terms of exporting/sharing time and/or version instances of, for example, storage images, files, messages, mailboxes, transactions or database instances.

It is therefore likely that no single CDP approach will satisfy the maze of attendant expectations, requirements and capabilities. Thus, the likely solution will comprise a combination of CDP solutions that collectively address the requirements through individual capability strengths.

8.4 Conclusion

CDP is an emerging technology that enables enterprises to increase their ability to provide sustained application availability through enhanced data recovery operational capabilities. It usually involves a variety of hardware technologies within a tiered storage environment as well as awareness of ILM's role in ensuring enterprise data's accessibility. It is important to recognise that CDP improves and expands rather than supersedes the data protection spectrum that legacy data protection and disaster recovery services already provide to enterprise users.

Finally, because of the multiplicity of products involved, it is important to select a CDP solution that minimises the number of interfaces data administrators that must master, preferably one integrated within a familiar and existing operational enterprise protection and recovery solution. Otherwise, the sheer number of point-product interfaces involved will rapidly transform simplified environments into ones that are vastly more complex.

9

What is the Cost of an IT Outage?

9.1 Failure is not an option

If CIOs are going to confront the dilemma of dealing with the convergence of IT and the business, then their first step is to ensure they have a robust Enterprise Storage Management system in place. What are the factors to take into account when selecting a storage vendor? Is it price, integration possibilities, virtualisation, scalability?

Simply put, the most important criterium must be the ability to manage the data processing, manipulation, management and transformation capabilities of the storage systems. The future of storage is a company's lifeline to its operation and customer base. Any vendor must have the vision and technological road map to look to the future as well as come up with a solution for today. Technical support, consulting integrity and seamless integration with existing hardware and software used are critical and mandatory. What is the cost of this? The millions saved by ensuring the operation is up and running 24×7, it mean that is more important to discuss the cost savings.

Prolific costs have been banded around over the past few years, showing this loss and that, for minutes, days and weeks of outage for differing vertical business sectors. And yet, it seems no one formula fits all. Gurus spend time writing about how difficult

---

Data Lifecycles: Managing Data for Strategic Advantage   Roger Reid, Gareth Fraser-King and W. David Schwaderer   © 2007 VERITAS Software Corporation. All rights reserved.

it is to write about and are unable to present a set of rules to derive the impact of downtime to an organisation. Granted, calculating an organisation's total financial loss due to network and system downtime is not a simple undertaking. There are many tangible and intangible impacts that must be factored into the equation.

The simple equation is easy and most of us can do it in our heads. A six-hour outage on a 24×7 €36 million per annum direct fulfilment system is:

$$(\text{€36 million}/(365 \times 24 = 8760 \text{ online hours per year}))$$

$$\times \text{ by } 6 \text{ hours} = \text{€24\,657.00.}$$

But that is a superficial calculation. Lost revenue is only the most obvious, most visible and easily identified cost of downtime, and the calculation above is a reasonable ballpark figure of that loss (notwithstanding sale fluctuations, calendar and time). But this simple, grossly inadequate calculation only lightly touches the real organisational costs. As organisations become more interdependent across business units and extend their supply chains, downtime impact escalates rapidly.

In fact, usually all aspects of an organisation's operation, production and development together with its support functions are affected by an outage, as is the outward appearance of the organisation to its existing and potential customers. Unreliability is one of the most detrimental characteristics to an organisation and that's exactly what an outage appears to be to the outside world – mitigating circumstances are irrelevant from the customer's point of view. This includes internal as well as external customers, potential customers, suppliers, government agencies and the competition within the vertical market sector. Any weakness the competition notes may well be used as a competitive advantage.

Simply put, the whole business is affected. To assess the financial impact of any outage accurately an organisation must consider all the aspects that are in use at any time within that business. The list is horribly exhaustive, including both tangible and intangible costs.

9.1.1 Tangible costs

In order to build a meaningful Return on investment (ROI) tool that calculates true downtime costs requires measuring all outage aspects. Tangible costs include all those aspects that are easily identified and measured in terms of hard cash. Tangible costs include lost revenue, the cost of wages for idled workers, remedial labour costs, vendor onsite costs, share depreciation, lost inventory, marketing costs, bank fees, late penalties and legal costs.

Here's a summary of the tangible costs:

- lost revenues;
- unemployed employees;
- loss of productivity;
- fines and penalties;
- legal costs;
- employees working on the outage rather than their day job;
- additional vendor technical support/consulting/technical engineer onsite costs (unless within a fix support contract).

The most evident downtime cost is lost revenue. If an organisation cannot process customers, it is effectively closed and cannot conduct business. Electronic commerce exacerbates the situation by amplifying the problem, as transactions are entirely dependent on system availability. One way to estimate the revenue lost due to a downtime event is to look at normal hourly sales and then multiply that figure by the number of hours of downtime as illustrated above – however, this is only one component of a larger equation and, when used alone, seriously underestimates the true loss to the organisation.

Employees are still paid during a system outage, and so are workplace space overheads. Other employees may still perform work, but their output is likely to be of lesser value (as we become more dependent on our systems we struggle to find any value away from our PCs). Projecting the number of labour hours that are lost in a future outage is difficult, but analyses of past downtime events can help provide a good estimate.

The most common way to calculate the cost of lost productivity is to first take an average of the hourly salary, benefits and overhead costs for the affected group. Then, multiply that figure by the number of hours of downtime. Unfortunately this is a very conservative cost estimate because most organisations run their business for profit, so employees contribute considerably more in value than the overhead cost involved in keeping them working. Repeating this calculation for each department and employee classification may prove important because the value of workers, and the degree to which they depend on systems, varies. Grouping similar employees into like categories may simplify this particular analysis.

Recovery labour requires an additional calculation. The employees that are idle due to a system failure are probably the same employees who resume work and recover the system via data entry. They not only have to perform their normal day-to-day activities, but they must also re-enter any data lost due to the system crash, or enter new data that was handwritten during the system outage. This means additional work, frequently on overtime at an increased hourly rate.

Some organisations work under contracts that include late delivery penalties. If system downtime causes a missed contract delivery date, the penalties incurred must be included in the calculation. Furthermore, organisations must file financial statements for tax purposes, and public organisations must comply with additional filing requirements. Any late-filing penalties incurred due to system outages also must be accounted for as downtime costs.

Depending on the nature of the affected systems, the legal costs associated with downtime can be significant, particularly in today's litigious society. In some European countries, if two organisations form a business partnership in which one organisation's ability to conduct business is dependent on the availability of the other organisation's systems, then, depending on the legal structure of the partnership, the second organisation may be liable to the first for profits lost during significant downtime events. If the lack of system availability results in the shipment of defective products, the organisation may incur product-liability costs.

The legal environment is complex. Any legal costs incurred as a result of past downtime should be used as an initial estimate of future liabilities. However, because each case is likely to be unique, legal advice should be taken to explore any other risks the organisation may face. For example, in some countries the management and the board are responsible for the share price; when there is a significant drop in share price shareholders may initiate joint civil legal action if they believe that the board were negligent in protecting vital assets.

Other costs inevitably arise during a downtime event. For example, a perishable-goods producer may have to dispose of spoiled goods, or a manufacturer may incur set-up costs to restart a stopped assembly line. It is impossible to list all potential costs that fall into this catch-all category, as many are specific to the affected organisation and its particular environment.

9.1.2 Intangible costs

Intangible costs are ones that tip the balance for data protection, high availability, storage management and disaster recovery strategies. These costs are not easily measured. Examples include lost opportunities, employee retention, share value, goodwill and brand damage, as well as remedial expenses.

Here's a fuller list of intangible costs:

- potential lost revenue;
- loss of contact data;
- inventory data, system and data recovery costs;
- failed Service Level Agreements;
- lost opportunities;
- lost potential customers;
- loss of existing customer loyalty;
- reputation – brand damage;
- goodwill;
- share depreciation;
- loss of supplier faith.

The crucial intangible cost area has only recently gained acknowledgement within most enterprise organisations. Now, the Gartner Group estimates that 80 % of all organisations calculating downtime include intangible costs in their business impact analysis calculations. Essentially, all organisations now acknowledge this is a critical area for outage impact measurement.

When a network outage prevents customers and prospective customers from dealing with an organisation, some do not try again but buy from the competition instead. A portion of these potential customers would have otherwise become loyal customers if they only hadn't been impeded by an unavailable system. An organisation therefore loses the immediate purchase of a potential customer as well as all the purchases that potential customer would have made over the life of the business relationship.

Consider a prospective customer who, in addition to an initial purchase, would have become a loyal customer and made one purchase every year thereafter for 20 years, each of which generated a profit of €1000. When calculating the value of this purchase stream, it is important to discount the future revenue to recognise that a euro received today, is worth more than a euro received tomorrow.

At a 5% interest rate, today's value for all the sales that would have accrued from this loyal customer would be about nine and a half times the value of the first purchase. If an organisation loses 1000 such customers due to frustration over system outages, its loss would total nearly €9.5 million. Clearly, an enterprise doesn't have to turn that income amount away too often to go out of business altogether. The assessment can be also calculated by existing customer revenue versus new business.

Intangible costs calculations should also consider the cost of replacing employees who may have become frustrated and quit frustrated by repetitive system downtime. Frequently sitting idle or redoing work can prove aggravating and lowered morale increases staffing turnover, raising hiring and training costs. Because it is nearly impossible to understand why an employee leaves an organisation, assessing the financial impact of a lost employee through outage frustration is difficult. But employees generally leave for much the same reasons that customers do not remain loyal. So, employee outage impact may be assessed in a similar way as customer loyalty.

In the late 1980s, a customer satisfaction guru (W. Edward Deming) quoted the 1-10-100 rule that said:

> *For every dollar your company might spend on preventing a quality problem, it will spend ten to inspect and correct the mistake after it occurs. In the worst case, the quality failure goes unanswered or unnoticed until after your customer has taken delivery. To fix the problem at this stage, you will probably pay about 100 times what you would have paid to prevent it happening at all.*

Although Deming originally applied this rule to manufacturing industries, it also applies to service industries and Federal Express used it in the early 1990s. Although there is little evidence to back this up, IT system reliance is now so great, perhaps a more applicable rule is a 1-100-1000 estimate.

Large organisations spend millions of euros building brands and developing and protecting their images. Repeated downtime may harm an organisation's image over and above the effect of any single downtime event. The compound impact could be felt in consumer confidence, sales, share value and reputation. If major downtime causes a loss in brand appeal and consumer confidence, an expensive corporate-image and marketing campaign may be necessary to repair the damage.

Likewise, if a large number of customers switched to the competition during an outage, the organisation may also need to initiate a costly promotional campaign, possibly including significant price discounts, to win them back. All expenses incurred to resolve these problems should be elements in the downtime cost calculation.

9.2 Finding the elusive ROI

Return on investment (ROI) is the critical metric in any investment decision, especially in the current economic climate. The expected benefits of a proposed project must exceed that project's anticipated costs by enough to justify the inherent risks. Again, this must now be achieved in relatively short order. With IT hardware shelf life becoming increasingly shorter, it is commensurately harder to write off any IT spent over a five or even three year period. So, ROI needs to be validated within six months to a year rather than over several years.

Because of the current financial climate, finite capital resources may still shelve a project if other investments promise to deliver better returns. A proposed managed data protection project will only proceed if a rational comparison of the projected costs and benefits can justify the capital spent.

Most IT application project benefits are a direct result of new application functionality. Most applications are designed to increase revenues, while systems that automate functions and previously paper-intensive processes, reduce costs. The benefit logic is reversed, however, for a managed data protection project, since the focus is not on what is gained by implementing the project, but what is lost by not implementing it. What CIOs consider is not what is gained, but what is potentially lost if the project does not proceed. Therefore, missing or ignoring costs when conducting a cost/benefit analysis for a data protection project often results in a gross underestimation of the project's potential benefits.

As always, the bottom line is profitability. Every business should manage data within regulatory requirements to the most appropriate level for its unique operational, marketing, competitive and financial situation. Finding the optimal level of data management investment requires considerable effort. But taking the time to investigate and understand the interdependencies certainly helps to quantify any potential downtime impact better. This is an essential first step in the quest for a data management programme.

Organisations need to look at a number of areas of the business to ascertain the final damage that any outage my cause. Not least, it all starts with the amount of revenue the organisation loses if its systems were down. If you take the average (taking into account the peaks and troughs of 24×7 transactions – it may not be that as many transactions take place at 2 am as at 2 pm) revenue per hour and multiply by the number of hours it would take to recover from the disaster, you will get a rough cost of lost productivity. Add the payroll, taxes, benefits and overtime for recovery. Then multiply by all affected employees. You can then identify the value of IT employee productivity lost during the outage by using the same process for all the IT engaged to rebuild and bring the IT systems back online. How much accounting information will be lost, and how much will it cost to recover or rebalance manufacturing processes? This will depend on the type of business you are in, of course, and will have a significant impact in the cost of the outage. What are the fees, fines, compensatory payments, breach of contract payments, regulatory fines, late-payment fees, solicitor fees, health and safety, or liability exposure liabilities you will have to pay? And finally, what steps will you have to take to recover your sales and marketing position following the outage around revenues, lost customer loyalty, reputation and/or goodwill?

9.3   Building a robust and resilient infrastructure

Given the damage that an outage can cause it is obvious that managing your data and its lifecycles also requires a robust architecture and a business continuity strategy. Inevitably, as companies seek new ways to beat their competition, they eventually examine their internal operations. A company's success revolves around information technology, whether directly or indirectly related to business processes. Companies' revenue stream, and hence their survival, now entirely depends on their business data and information. As we have seen, the growing complexity of information technology and IT infrastructures leads many organisations to react by installing additional amounts of server and storage capacity to counter the problems of storage and data bottlenecks. And, as we have also seen, the effect is that the extra capacity is usually dedicated to specific servers that are unavailable to help in other areas when needed.

As enterprise organisations seek to grow and prosper, they need to ensure the secure and reliable availability of IT services to keep business up and running at all times. And that requires building an infrastructure that is flexible enough to respond to a changing IT environment, but resilient enough to withstand disruptions. Data, information and the means to act upon that information, remains one of the most significant business assets. Instead of treating IT as a by-product of the business, organisations need to think carefully about how to protect their IT infrastructure, as well as access and share data, and ensure its availability to current and future applications. Managing information is much more than simply putting another storage device on another server and making sure it is protected from viruses.

As critical components of every business application, both security and storage need the same careful design as networks and systems – creating the concept of a robust architecture. The IT architecture is more than simply the topology of how you connect clients to applications to databases, servers and storage, and more than how to protect against malicious threat; the architecture must include the people, processes, hardware and software that support data in the organisation. To face the challenges of a changing business environment, organisations need to create an architecture that can adapt to change quickly and efficiently. Unfortunately, since critical tools aren't always interoperable and IT operations and security functions often have conflicting priorities; in these cases, some solutions can create more complexity, and more problems, than they solve.

9.3.1 Five interrelated steps to building a resilient infrastructure

Companies need to:

• set a baseline by evaluating their assets and security systems, assess current procedures against risks and set goals and policies for desired levels of availability and security

• identify and analyse threats and create a plan to meet them;

• deploy proactive safeguards against potential threats, enabling it to protect assets and recover from disruptions quickly;

• remediate threats quickly and implement long-term solutions for vulnerabilities, including regularly updated policies and new certification standards for compliance;

• proactively monitor its security, information management and storage systems to provide a clear view of the entire infrastructure and allow a rapid response to any disruption.

Technology alone cannot create a resilient infrastructure. You also need to combine the right people, processes and tools which deliver an optimal solution that preserves the integrity of business-critical operations. A broad range of solutions are available to make a resilient infrastructure a reality for any organisation. These solutions establish processes to increase the efficiencies of your IT security and operations teams and align them toward the same goal: to build an infrastructure that ensures the security and availability of information systems and preserves business continuity.

9.3.2 Disaster recovery concepts and technologies

There are two concepts that you can use to describe the recovery of data, servers and applications (see Figure 9.1):

• Recovery Point Objective (RPO) – how much data loss an organisation can afford to lose if a disaster occurs.

• Recovery Time Objective (RTO) – the time it takes, following a disaster, for your applications and the data associated with those applications to get backup and running.

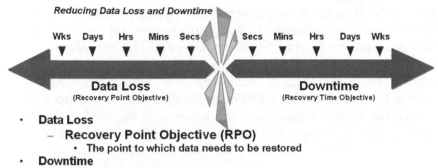

- **Data Loss**
  - **Recovery Point Objective (RPO)**
    - **The point to which data needs to be restored**
- **Downtime**
  - Recovery Time Objective (RTO)
    - **The time by which data and systems need to be recovered and restored**

Figure 9.1 Technologies for IT business continuity

The greater the need for a speedy RPO, the higher cost associated with it. In the same way, the sooner the data and applications are needed to be operational, the higher the cost.

There are three technologies for RPO:

1. vaulting;
2. disk backup;
3. replication.

There are four technologies for RTO:

1. restoring data from tape;
2. restoring data from disc;
3. clustering;
4. replication.

The RPO/RTO must be assessed and based on an individual application since each application may have different needs. The easiest way to conduct RPO/RTO assessments would be to consider that every application is critical to the business – every application requires the most up to date data as well as the applications back online in the shortest space of time. But for most organisations, the costs are simply prohibitive. The more technology you throw at disaster recovery, the greater the cost associated with that project.

A recent project identifies 57 different applications that the individual businesses considered to be 'critical'. When analysed it turned out that, in fact, there were only two applications that were fundamental to the business. Other applications were provided for based on their relevance to the requirement the organisation had for business continuity. The solutions ranged from a platinum level disaster recovery solution; (including 24 × 7 scheduled availability with a RTO of 2 hrs and a RPO of 0 hrs) through to a copper level of disaster recovery with working hours of scheduled availability with a RTO of 10 days and a RPO of 1 day (see Figure 9.2).

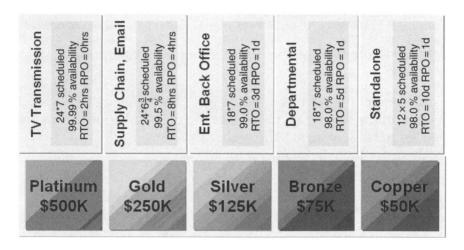

Figure 9.2 Five classes of service – application owner bears the cost of the service

9.3.3 Disaster tolerance

In order to review the possible business impact of an IT outage you must first consider an organisation's disaster tolerance and second provide the ROI for the investment in IT management systems that will help the organisation manage the impact of a disaster. Disaster tolerance looks at catastrophes from two perspectives: i) disaster avoidance and disaster recovery; or ii) high availability of IT systems and how they relate to the broader concepts of business continuity. Disaster tolerance could be considered a special case of high availability or the most robust, logical extremes of high availability. High availability is a term that covers a wide range of issues from the level of system performance to keeping the door open 24 hours a day. Regardless of the way it is defined, however, disaster tolerance addresses a specific and concrete problem – how to survive a catastrophic event that could, at its worst, destroy an organisation's ability to function at all! Disaster tolerance can only be defined for a specific organisation when we include the vertical, geographical and demographical characteristics of the business. Businesses like manufacturers have tangible production and inventory issues and others like banks and insurance companies have a huge amount of customer data, information and business intelligence to manage within the life-cycle. If a retail business is halted then the immediate requirement is to protect the critical information resources such as bills of material; for a bank it will be to ensure the support of customer transactions.

There are some basic questions an organisation can ask itself in order to ascertain how tolerant IT may be to a disaster:

1. Do you have a disaster recovery plan?
2. Is your disaster recovery plan part of a larger business continuity plan?
3. Do all parties – senior management, operations, legal, IT, security and facilities – know about the disaster recovery plan?
4. What technology do you currently use for high availability, disaster recovery, and business continuity? Do you use

   – backup?
   – replication?
   – clustering?

5. Do you use power backup and power fluctuation systems to protect your business?
6. Can you accurately say how quickly you can recover from a disaster?
7. Do you test your disaster recovery plan weekly or monthly?
8. Do you know your recovery point objective for all customer-facing systems?
9. Do you know your recovery point objective for all partner-facing systems?
10. Do you know your recovery point objective for all internal-facing systems?

11. Do you store backups of your data (media tapes) offsite?
12. Do you protect laptops?
13. Do you protect desktops?
14. Have you measured how much one hour of downtime would cost you?

9.4   Conclusion – Analysing business impact

When we talk about the impact that system failure can have on a business we usually think of loss of revenue during downtime. However, the impact is far greater than simply the inability to carry out transactions. The true effect to a business should be measured in terms of operations, financials, regulatory and legal implications. If a business is forced to operate without its systems for a significant period of time the day-to-day information we are used to extracting at the touch of a button will require tedious research and manual labour. Transactions that would normally take place automatically will cease, staff will be unemployed without these systems, personnel and operating regulations will be more difficult to adhere to and subsequently the efficiency of a business will suffer and management will be faced with making vital decisions to carry the operation through the crisis.

If a business's systems are interrupted it will experience some serious financial losses, in many cases these can be fatal; these losses are no longer covered by insurance. Should any of the fundamental functions within a business be disrupted then inevitably the lack of cash flow will prevent the company from deploying capital to keep the business running. The ultimate risk is that credit ratings suffer, shares go down, the business is unable to fulfil the legal requirements of suppliers contracts, customers lose faith and may never return, and significant costs can be associated with restoring servers and systems.

So when looking at the most appropriate data protection solution for your business, the business contingency planner will need to identify the critical functions within the business and estimate the business loss for an outage and the effect that time has on the impact. Once you have determined the impact of an incident on a business function, you can determine the recommended recovery timeframe for the function.

9.4.1   Identifying critical functions

The main areas of assessment should be:

• reviewing business expectations and service levels;
• reviewing existing business continuity measures;
• conducting a vulnerability study of existing infrastructure;
• reviewing operational policies and procedures;
• assessing recovery capabilities;
• examining options for reducing risk exposure.

Critical applications can be identified in a number of ways:

• list all functions performed;
• interview business functional managers;
• conduct surveys with employees;
• use a questionnaire with the functional managers.

Once the critical and necessary business functions have been identified, the next step is to establish the resources that are required to continue to perform those functions. Ensuring that you have the correct level of support for these technologies is imperative as well as ensuring that the solution you deploy is implemented and managed effectively.

Training your staff has never become more important. What is the cost of downtime against the cost of continuing professional education? When spending a good deal of time and effort in implementing a data storage strategy, why jeopardise that investment by not ensuring its on-going management and strength? The cost of downtime in the economy, in which we now live, means that any system outage can have disastrous effects on businesses where the cost of downtime accelerates as the outage increases in duration. Giving your IT staff the skills necessary to ensure your data storage strategy is fully effective and protecting your business helps businesses achieve maximum return on investment, helps reduce capital and expenditure by decreasing operating costs and increasing productivity, reduces technical infrastructure costs, increases the effectiveness and retention of IT staff and increases technological competitiveness by improving effective access to business critical information.

In the longer term, the business will need to be restored to its original performance. Identification of all resources required to support the function is required to facilitate the longer-term recovery. Even a small business can lose thousands of pounds an hour due to network downtime. Your business needs a comprehensive solution to maximise your performance and productivity in your investment in technology. Today's operations are highly interactive; each part of the workflow relies on another process or system. Your network is the tool your systems use to communicate with your customers, suppliers and employees. When you have network downtime, your business is losing money. Productivity suffers and your reputation can be damaged. Directly or indirectly, a system outage can cause a loss of sales for your business. If you are engaged in e-business, with your doors effectively shut, customers will turn elsewhere.

10

Business Impact

*Humans are symbol-making creatures. We communicate by symbols –
growls and grunts, hand signals, and drawings painted on cave walls in
prehistoric times. Later we developed languages, associating sounds with
ideas. Eventually Homo Sapiens developed writing, perhaps first symbols
scratched on rocks, then written more permanently on tablets, papyrus, and
paper.*

Bell Celebrates 50 Years of Information Theory

10.1 Business impact

Ever wonder why so many strategic storage plans wash away
never to be heard again? The answer: companies are being stunned,
paralysed and over-assessed without specific recommendations
to help them accomplish goals. They have seen it all – from
assessments to analysis that never produce viable solutions. What
they receive is basically lip service promising future data analysis nirvana. However, what they need are strategic plans that
span the business and can be viewed in a clear, straightforward
manner. Any strategic plan should contain visual and written
documentation that emphasises various business characteristics
and concepts for how the company should channel activities
inside and outside the company. Information lifecycle management (ILM) should be a riveting concern to the entire company and
not only to the IT department because it can significantly impact
critical business systems. Remember, the compliance discussion

Data Lifecycles: Managing Data for Strategic Advantage   Roger Reid, Gareth Fraser-King and
W. David Schwaderer   © 2007 VERITAS Software Corporation. All rights reserved.

that addressed regulatory compliance measures for various sectors such as financial, health, telecommunications and government entities? This topic should not be taken lightly by any organisation or employee. Therefore, to develop such a comprehensive plan, all departments should participate at some level.

Before continuing, beware. This chapter is not a deep dive into the typical phases of *Business Impact Analysis* as it relates to *Business Continuity*. Instead, it uses some of those techniques coupled with the *Unified Modelling Language* that explores a mapping methodology, which strongly suggests that the approach provides a useful integration tool for modelling data, storage/computing resources and business process flows. This approach is beneficial for current and future DLM and ILM projects that can produce positive business impact. Moreover, it touches upon the criticality of adopting new technologies to position the company competitively.

10.1.1 Business impact analysis

Previous chapters discussed the use of the *Business Infrastructure Mapping Analysis* (BIMA) to assist this process. Using the process helps to communicate an interrelated plan that management can communicate unambiguously throughout the company, potentially resulting in new ways to use recommended software tools, methodologies and processes to reduce the complexity in managing accelerating data rates. Moreover, it assists companies in meeting various business service level agreements that entail government, corporate and regulatory compliance measures presently transforming business conduct across the globe.

That said, you might probably be asking yourself, 'How does this impact my business?' Below is a summarised list of reasons why you should develop lifecycle models of your business for data:

- helps identify business applications, hardware and storage which communicates lifecycle architecture designs quickly and effectively;
- makes models actionable against associated technology;
- helps understanding your business and its technology-linked processes;
- helps identify how change affects your business;

- helps identify strengths, weaknesses and areas that need change or optimisation;
- assists divergent business units and IT teams to work and communicate more effectively, potentially saving organisation time and reducing costs;
- helps identify the organisation's willingness to adopt technologies and invest in technology education to stay competitive.

All the above reasons play an integral role in whether you or your company can be competitive in the new market. A few questions that you should ask are:

- In case of business climate change, are employees aware of and trained in carrying out different roles and responsibilities?
- Are business processes, systems, applications and data stores well documented to provide an effective visual story of the enterprise?

And, since this is a time of intense financial review, being competitive means staying in business. As a central starting point for pursuing an effective lifecycle strategy, previous discussion suggested a series of processes and ideas that could help achieve an enterprise's vision and mission objectives. Just as a thesis represents a written proposition relating to a particular subject, consider using a thesis in this context. Having a clearly defined proposition on how an enterprise should management the data lifecycle is an essential step.

Thoughtful corporate citizenship requires carefully devising actionable plans or propositions for controlling costs and insuring data integrity at all levels. Therefore, goal and mission development is the initial step in developing the storage management profile. Next, consider the selection process and how to apply it to specific product categories. Simultaneously give an overview of how the cost is associated with its infrastructure adoption. There are various technologies that could potentially couple the process and methodology to assist in the overall mission and goals, but are they worth it?

As discussed previously, all departments should participate in this effort by examining and documenting their systems, applications and processes. This helps identify necessary data needs. Examining this further reveals that it is a complex task, requiring coordinated efforts between departments to capture necessary business needs. Once completed, the strategy can further evolve by identifying additional technologies to assist in completing the vision and business objectives. Some of the questions that might surface are:

- Where should we start?
- What tools should be used?
- Should this task be taken at the entire infrastructure or in sections?

With shared-data applications and storage, the responsibility to ensure availability to current and future applications requires a common technique or approach to assist the documentation process. As critical components of every business application, both security and storage need the same careful thoughtful consideration as networks and systems – creating the concept of an integrated utility architecture. The IT architecture is more than simply the topology of how to connect clients to applications and databases, servers and storage, and how to protect it against malicious threat; it must include the people, processes, hardware and software that support data in the organisation.

To face the challenges of a changing business environment, organisations need clear documentation in the form of infrastructure models so the business can adapt and change quickly and efficiently. Unfortunately, since critical tools do not always use standard notation, some functions often have conflicting meanings. In most cases, some solutions can unfortunately create more complexity and problems than they solve.

Therefore, in order to meet the objectives, a language such as the *Uniformed Modelling Language* (UML) is a great tool to begin the process. Or, other models also exist; they would likely prove useful in mapping lifecycle project requirements. Just as music, electrical engineering and architecture use standard tools in documentation, the same is true for UML. Here, UML is a 'common' language that helps business analysts, software developers and others to understand the business requirements. IT specialists could use it as a tool to describe DLM and ILM. By carefully modelling the business, UML can help identify aspects of the company that were previously not understood. Consider the simple model depicted in Figure 10.1 (admittedly not in UML standard language, but a model nonetheless) as an example of conducting the ILM process.

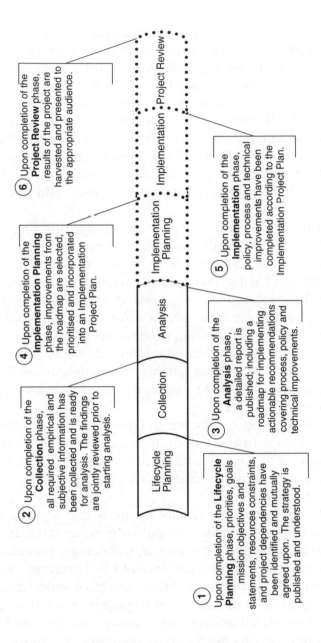

Figure 10.1 Information Lifecycle Process Model

① Upon completion of the **Lifecycle Planning** phase, priorities, goals mission objectives and statements, resources constraints, and project dependencies have been identified and mutually agreed upon. The strategy is published and understood.

② Upon completion of the **Collection** phase, all required empirical and subjective information has been collected and is ready for analysis. The findings are jointly reviewed prior to starting analysis.

③ Upon completion of the **Analysis** phase, a detailed report is published; including a roadmap for implementing actionable recommendations covering process, policy and technical improvements.

④ Upon completion of the **Implementation Planning** phase, improvements from the roadmap are selected, prioritised and incorporated into an Implementation Project Plan.

⑤ Upon completion of the **Implementation** phase, policy, process and technical improvements have been completed according to the Implementation Project Plan.

⑥ Upon completion of the **Project Review** phase, results of the project are harvested and presented to the appropriate audience.

Lifecycle Planning · Collection · Analysis · Implementation Planning · Implementation · Project Review

The process model shows various lifecycle process 'actions' or 'activities'. With the above visual description, divergent groups could understand the engagement process for a lifecycle project. Even though it should be depicted in a standard notation, the model clearly shows what steps are needed to achieve the associated goals. Furthermore, the figure begins with the lifecycle planning stage where it clearly articulates the need to plan data categorisation properly. Furthermore, it highlights the importance of preparation to provide a proper framework in which certain technology techniques or actions could be used and applied to the data.

Consider the categorisation process or 'storage profiling' shown in Figure 10.1. This is basically a set of pre-defined criteria in which the data can be further categorised. Then, prescribed conditions can be applied to it and action taken using various technologies. Using models for this process helps understand where an enterprise is today and where it wants to be. In most cases, to avoid being overwhelmed with the modelling process, the examination should be undertaken at a reduced scope within a division, sub-division or at a more tactical level. Most notably, the models for the lifecycle project are normally selected to achieve a specific business goal or eliminate a specific inefficiency.

Another visual model is shown in Figure 10.2, which carefully depicts a resource with a certain configuration. In addition, it could visually depict a storage asset and attributes. At the bottom, it represents a business service model for various service levels associated with the assets.

Just as technology alone cannot create sound storage infrastructures, models alone cannot execute actionable tasks. It is also necessary to combine the right people, technology, processes and modelling tools to deliver optimal solutions that preserve business-critical operational integrity. A broad range of modelling solutions exists to assist in documenting organisational infrastructure realities. Moreover, modelling can broadly establish, and effectively communicate, processes across organisations, languages and countries – thereby increasing the efficiencies of IT and business teams, and aligning them toward the same goal: To document and build an infrastructure that ensures the data, security, availability and management of the lifecycle of information systems.

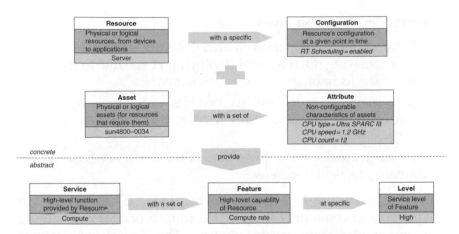

Figure 10.2 Resource with a certain configuration. (Tactical point: Business models help people understand a business. Very simple visual models can be great focus points and are critical in managing a data lifecycle.)

10.1.2 Cost versus adoption

There have been numerous software industry studies on the use of the various metrics including:

- Return on Investment (ROI);
- Total Cost of Ownership (TCO);
- Return on Assets (ROA).

These metrics have successfully identified the usefulness of either an enterprise software application or hardware. However, according to Ross Mayfield, an entrepreneur and technology marketing executive: 'There has been a significant shift from using ROI to TCO to justify technology purchasing.' That said, although the three metrics can support a certain hardware and software acquisition business case, many people are unaware of how the three metrics *contextualise* the investment to assess technology and its associated risks. This is essential in understanding the importance of implementing lifecycle processes, methodologies and technologies in any environment as well as the introduction to utility computing paradigms.

Further examination reveals that

- ROI calculates the expected financial return, discounted for the risk of achieving the return, of deploying IT relative to the direct costs of the technology and its deployment.

- TCO contextualises technology benefits to business operations by assessing direct and indirect costs that create operational expenses.

- ROA contextualises the benefits within the asset base of the company, or capital expenses.

However, using the *Technology Adoption Lifecycle* can provide an adaptable and commonly understood framework for considering the timing of certain business cases. The *Technology Adoption Lifecycle*, coupled with the *Technology Valuation Lifecycle* (TVL), describes and outlines the various segments and the psycho-demographic profile of a technology prospect.

To begin, the categorised and described technology adoption segments are:

- Early Adopters;
- Early Majority;
- Late Majority;
- Laggards.

You may wonder what Early Adopters, Early Majority, Late Majority and Laggards are: a very good question, which we address in the next few sections using quotes from the White Paper located at the following web address: http://radio.weblogs.com/0114726/ whitepapers/Timing Your Business Case-Ross Mayfield.pdf.

10.1.2.1 Early adopters

Early Adaptors are

> . . . *Visionaries [who] acquire technology as a change agent, for competitive differentiation. ROI is a sufficient measure for this objective, because the project is usually focused on creating new strategic advantages outside the focus of core operations. ROI is direct competition. Visionaries are risk-tolerant in that they will invest time to learn new technologies and will accept ROI models based upon reasonable assumptions. Since investment objective is as a change agent, the operational and capital expenditure costs for deployment are less of a risk factor than achieving basic competitive advantage with basic returns on invested capital.*

10.1.2.2 Early majority

The Early Majority are

. . . Pragmatists [who] acquire technology for productivity improvements. TCO contextualises benefits with the operations of the company, as measured by operating expenditures. TCO informs decisions on technologies of similar benefits. ROI is used to support the argument, but will not be the metric focus. This segment is less risk-tolerant than the aforementioned, and requires references to offset risks. This means that TCO models cannot be pure theory. They require actual case studies from comparable customers or deployments to stress test results. Their strong reference requirements may also require third party validation (e.g. from an Industry Analyst) of the model or cases. Since the investment objective is to improve productivity, the technology is deployed within the context of operations, putting greater focus on operational expenditures created by deployment.

10.1.2.3 Late majority

The Late Majority are

. . . Conservatives [who] purchase technology to maintain competitive equivalence with the mainstream market by maximising asset utilisation. At this point in the lifecycle technologies are well tested and have proven their value. In addition, competitive pressures from competing providers reduce the price as these once hot technologies have become commodities. ROA contextualises benefits within the asset base of the company. ROI and TCO are used to support the business case, but are not the metric of focus. This segment is even less risk-tolerant and will not invest to learn how to use technologies. Since investment objective is maximising asset utilisation, the focus is on the price and price risk of the commodity.

10.1.2.4 Laggard

Laggards

. . . purchase technology, if at all, when it is deeply embedded in other products and do not factor valuation of the technology into a purchasing decision.

Using this framework – the *Technology Valuation Lifecycle* – Early Adopters willingly risk investments to learn new technologies, even when a technology is largely unproven. Moreover, it suggests that the framework provides Early Adopters with a *Strategic Advantage* coupled with an opportunity for financial reward. This alone captures the segment's competitive differentiation. The Late

Majority, by contrast, are unwilling to invest in learning and demand market-proven technologies because it is less risk-tolerant. Figure 10.3 depicts how the Technology Adoption Lifecycle maps to a Technology Valuation Lifecycle.

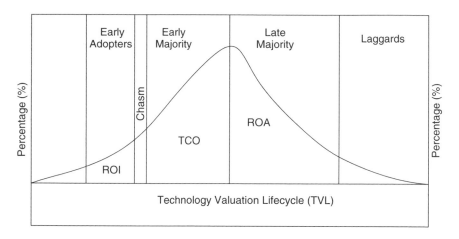

Figure 10.3 Technology Valuation Lifecycle

That said, with such a visual description, it is clear the Technology Lifecycle framework can be a useful tool in determining whether certain lifecycle technologies benefit a company and provide overall advantage to the company.

Meanwhile, business scenarios continue evolving independently. Corporate mergers, restructuring, short-notice tactical purchase decisions and proprietary IT solutions, all make IT infrastructures more complex and difficult to manage. As adaptability inevitably declines, companies cannot react as quickly or cost-efficiently to changing market conditions. The effect is that many companies expend too much energy reactively instead of focusing on future-oriented, proactive solutions selection and development that can contend with local and global change.

As enterprise organisations seek to grow and prosper, they need to ensure the secure and reliable availability of IT services to keep business up and running at all times. And that requires building an infrastructure that is flexible enough to respond to a changing IT environment, but resilient enough to withstand disruptions. As we have repeated before: Data, information and the means to act upon that information remains one of the most significant business assets. Instead of treating IT as a by-product of the business, organisations need to think carefully about how to protect their IT infrastructure. From the preceding discussion, one can conclude that modelling is useful not only in documenting an infrastructure, but also as an essential part in the overall lifecycle management of data and information.

10.1.3 Service level agreements and quality of storage service

The first step in developing a storage thesis is to approach it tactically – focusing narrowly on the storage environment. The actual exploration process includes collecting various application storage requirements. If IT departments adopt a strategic lifecycle management approach, they explore the opportunity in more depth, searching for additional opportunities to provide cost savings. The models discussed earlier are great sources in determining SLAs and meeting quality standards. With the plethora of available industry information, it is unlikely to be necessary to have to convince colleagues about the need to develop lifecycle policies that provide different storage service levels mapped to storage tiers. Chapter 7 discussed tiered storage and how it could apply to various service levels that meet business requirements. Interestingly enough, it shouldn't necessarily be hard to convince business units to assist in providing their requirements.

Information Management is a critical component of your IT infrastructure and constitutes a large portion of the company's IT spending. To better understand, manage and justify additional investments in your storage infrastructure, various key resources should be involved. Each member would be selected by their job responsibilities as an *information broker* to assist in clearly defining service requirements. Inquiry questions can be from a list of structured questions or topics such as:

• applications and data subjects living on server(s);
• practices and procedures to manage the server(s);
• issues that exist today or threaten to impact negatively storage manageability, accessibility or capacity of the server(s).

The above list can aid and expedite data gathering. In addition, it is often prudent to request copies of specific documentation such as architectures, implementation or migration plans, data centre diagrams, etc. An interview process could fill in information gaps (if any) that the collective team identifies as a result of what is provided in the documentation. Note that interviews will not only provide answers to a particular subset of questions, but also expand on the answers in an effort to gain insight into IT infrastructure, architecture and processes.

10.2   The paradigm shift in the way IT does business

10.2.1   Aligning business with IT

Typically, capital expenditures are flat, but operational expenses continue growing. However, needs are broad and diverse. You have probably identified some unaddressed needs in your enterprise. However, after speaking with customers over the last few years, consultants have identified additional needs as well. Effectively addressing existing areas of need is important, not only in a tactical nature but also strategically. Nonetheless, the foundations of these needs are often the same and worth the identification challenge:

- capacity and performance planning for applications and storage;
- workflow to manage and fulfil IT services requests;
- simplified storage, server and application management through virtualisation;
- policy-based administration and enforcement;
- quality of Storage and Service Level Agreement enforcement across the enterprise;
- consolidated chargeback ('bill') across all IT services;
- attribute-based storage provisioning;
- hardware and software asset management across the enterprise;
- software distribution, maintenance, propagation and rollback;
- management of third-party products and services, including arrays and switches, applications and storage software vendors' point products.

The list might not be exhaustive, but it clearly shows the breadth of need. It also reveals the required breadth of people and/or groups. Stakeholders who are frequently involved in solving some of these needs often comprise three separate groups (see below).

The new paradigm shift is how to align the business with IT. Not aligning the business with IT will have a negative impact on the business one way or another. Outlined below are the three functional groups responsible for managing different aspects of IT service management or administration:

- **Business Manager**: Senior IT, or line of business staff focused on creation and delivery of business services.

- **IT Service Manager**: Architects or Senior IT Staff responsible for large IT deployments. Focused on availability and performance delivery: Multi-Domain.

- **Technology Administrator:** System Administrators, Database Administrators, etc. Focused on technical intricacies, less likely to be involved in the overall vision and mission.

These three diverse groups are responsible for continuing business impact. Managing data growth, defining the value and aligning the cost apportionment are vital factors and should be viewed strategically (Figure 10.4).

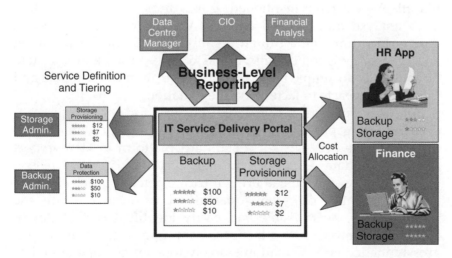

Figure 10.4 Aligning business with IT functions

10.2.2   Software consistency and agnostic support

To manage data, certain business rules (in fact, polices and constraints) must be observed. Business value is derived by having a consistent software layer used as an enterprise standard. Standardising the underlying foundation opens a wide range of new, inexpensive and flexible ways to build services for internal or external customers. Still, the vision is bigger. Data can flow across tiers, viewed and managed with tools and processes that are not only standard but also agnostic against the plethora of hardware that saturates the infrastructure. This provides the ability to reduce certain cycle times, improve decision-making, and reduce costs. Through a flexible management platform, business processes could seamlessly incorporate value-added features by removing manual steps that previously required weeks to accomplish. A responsive and adaptive business results when a business and its IT react quickly, aligned in goals and objectives. Now, how can your enterprise get there?

10.3   The Holy Grail: standard software platform

Prolific cost studies showing the cost of outages lasting hours, minutes, days and weeks for differing vertical business sectors have existed for years. Yet, it seems no single study approach fits all. As we have mentioned, gurus indicate how difficult it is to analyse outage costs and are unable to present rules to derive downtime impact to an organisation. Granted, calculating an organisation's total financial loss due to network and system downtime is no simple undertaking. There are many tangible and intangible impacts to factor into the equation.

But, have enterprises calculated the costs for not using a standard reporting platform? The next few sections provide helpful insights to the necessity for a standard platform to view, report and manage data. Data goes through cycles and for that reason reporting is essential in gathering intricate details of such changes.

Any calculation can seem superficial. Lost revenue is only the most obvious, most visible and easily identified cost of downtime when viewed from a business continuity view. This simple, grossly inadequate calculation scarcely touches genuine organisational costs. As organisations become more interdependent across business units and extend their supply chains, the impact of being unable to produce standard storage reports escalates rapidly.

In fact, all aspects of an organisation's operation, production and development together with their support functions usually experience the same outcome, as does the outward appearance of the organisation to existing and potential customers. Reporting unreliability is one of the most detrimental organisational characteristics – that's exactly what must not happen. Ineffective or ill purposed reporting for charge back levies can often appear as a major business fault to a customer. This includes internal as well as external customers, potential customers, suppliers, government agencies and the competition within the vertical market sector. Any weakness the competition notes may well be used as a competitive advantage. Simply put, the whole business is affected.

In Chapter 9 we quoted W. Edward Deming's 1-10-100 rule. The current financial climate means that finite capital resources may still shelve a project if other investments promise to deliver better returns. A proposed managed data protection project will only proceed if a rational comparison of the projected costs and benefits can justify the capital spent. So, how can you internally enhance your competitive visibility?

As companies seek new ways to best competition, they eventually examine their internal operations. A company's success revolves around information technology, whether directly or indirectly related to business process. Their revenue stream, hence their survival, now entirely depends on their business data and information. As we have seen, the growing complexity of information technology and IT infrastructures leads companies to seek out tools to give them an 'edge' – they search for software to give a detailed account for their storage assets and how it is being allocated across the enterprise. Each line of business has a stake in this. And, as we have seen, the effect is that the extra storage capacity is usually dedicated to specific servers that are unavailable to help in other areas when needed.

10.3.1 Business technology reporting and billing

With every move a company makes acquiring different technologies to address business cases, it continually captures more data and transforms it into information. The business needs a way to view the company's storage environment and distil it into meaningful reports, charts and graphs that help network, storage and business managers make tactical and strategic decisions about how best to manage their storage resources.

Due to the hardware plethora, the reporting platform has to be heterogeneous. It must effectively collect raw data from the various UNIX, Linux and Windows platforms. More importantly, it needs to correlate data by business unit, department, core business applications, etc. Here is just some of the information that a standard storage resource management (SRM) application should support:

- trending by server, volume, department, file types and applications;
- fastest growing servers, departments, volumes and users;
- key capacity information on storage devices;
- file information by

  - file type extension
  - files not accessed recently
  - duplicate files across volumes
  - largest files
  - largest users
  - largest subdirectories

- application specific storage information (Exchange, Oracle, SQL, Sybase);
- information for various departments and/or executives on corporate intranet sites;
- summary of network storage usage (which you can also use in management reports).

10.3.2 Smart storage resource management

Storage is often the most costly data centre component, yet often has the fewest tools and resources to track usage and utilisation. SRM involves both active and passive storage resource management. How can an enterprise manage these resources effectively if it doesn't know what's going on? With intelligent SRM products, enterprises gain a complete, 360-degree view of storage assets and how they are used. This allows them to plan for growth proactively while eliminating wasted resources.

Why do enterprises need SRM? In a nutshell, a company uses the tools to do the following:

- recover wasted storage;
- proactively detect storage problems;
- project future storage requirements;
- increase administrator efficiency.

Most lifecycle projects are focused on recovering wasted storage. Email and databases always utilise more and more storage. SRM can help by revealing how storage is used enabling wasted storage reclamation for ongoing and future projects. In addition, SRM can help consolidate reporting across enterprise servers and storage types such as SAN, NAS and DAS. Further, reporting summaries can provide information by server and volume. In a more granular view, SRM can gather information and provide reporting on usage by user and application(s).

10.3.3 Data forecasting and trending

A typical SRM tool shows usage trends and makes forward projections. It includes the ability to gather data both locally and remotely. This means enterprises can get baseline information without huge deployment costs. Consider the progress enterprises can make when they have information on historical usage trends and future usage projections. They will be able to perform predictive modelling to provide storage return-on investment analysis and departmental charge-back.

An integrated and intelligent reporting platform which combines the best of a SAN management tool and SRM would provide enterprises with the ability to manage the entire data path completely and effectively. In the bigger picture, if enterprises could discover and visualise the storage area network and then provide more granular reports, it would be well worth it. For example, enterprises could identify Microsoft Exchange resources, hosts and databases, while ensuring optimal storage network performance and availability. Here, SRM unveils storage utilisation by departments and users, and by predicting future usage trends.

With these two powerful tools, you have an entire SRM solution. If enterprises use SRM or plan to use it with mail servers, they can track files and storage utilisation by server. Armed with this information, enterprises can predict future storage capacity requirements, identify the users and departments with the most mail files and establish policies that maximise storage resource allocation. Using these two capabilities, enterprises can generate chargeback methodologies since they provide specific information on the use and capacity of each department's mail storage.

Effective storage infrastructure management via a combined set of enriched tools provides real, active, end-to-end storage resource management. With such SRM solutions, IT managers have what they need to manage their entire storage infrastructure more effectively. When looking for solutions, remember that a true solution provides real-time storage network management and monitoring, while providing detailed storage utilisation reporting. With such a total solution, IT managers are able to track hardware and software connectivity, performance and resource relationships. With intelligent resource management solutions, administrators can drill down into the organisation's storage utilisation, identify opportunities to clean disks containing unnecessary files,reallocate storage based on utilisation, understand and predict organisational usage trends and make better decisions regarding additional storage resources. As one would expect, such a solution pays for itself many times over in the time saved by users and network administrators. Moreover, they gain peace of mind from knowing that the organisation's storage resource investment is well managed.

It is common knowledge that this is just another element or component needed to manage the entire storage and data stack, albeit an important one. By bundling intelligent SRM capabilities to view important virtualisation objects such as volumes and file system technology components together, and integrating it with other technology elements – such as clustering, replication and backup – enterprises can demonstrate their ability to leverage technology and provide an additional enabling layer in their overall lifecycle vision.

10.3.4 Policy-based Administration

Normally policy creation is initiated by outlining initial storage requirements. Modelling requirements highlights the required actions. To accomplish managing the personification of data, it is desirable to have a solution that is policy-based and complies with regulatory compliance and corporate policies. In addition, it would necessarily support various archival functionality and features such as versioning, searching with indices, retrieval and recovery. In addition, it is also prudent to provide auditing and extraction elements. Having the ability to create policies and prioritise mission-critical data is one of the most important aspects of data lifecycle management. Without the means of setting these data delivery policies, all infrastructure data receives equal treatment, which is not what most enterprises would want. For example, inconsequential but large graphics files that are not often used would receive the same priority as an online transaction for a critical customer order. Both types of data would consume the same high price storage.

Another example, consider your company's marketing department. Marketing work often requires using large files, such as graphics-rich documents, presentations and sound files. But does marketing need to keep campaign files for three years? Do a dozen people in the department need to keep copies of those files? With an agnostic reporting tool, you can quickly and easily identify how storage is used and can set policies that free up storage resources, thereby lowering costs.

10.4 Summary

This chapter began with a discussion on the various business impacts of not having a strategic lifecycle management plan and why a strategic lifecycle management plan is important in achieving business goals. Enterprises can have a vision, but the mission and objectives should be carefully noted. Otherwise, enterprises have learned that the unfortunate consequences could be meeting their mission, but having drastic consequences by not understanding their requirements. Therefore, requirements should also be identified at the beginning of any project.

The chapter then discussed using models, how they could be used in a project and how important they are in communicating a vision, mission and mission objectives to a wide variety of people. Overall, the models should capture strategic structure interactions such as business assets used to fulfil the business functions and provide a common baseline for which enterprises can move forward with developing their lifecycle policies. As the infrastructure becomes larger, complexity becomes the largest competitor and managing this complexity will only worsen without modelling the interrelationships.

The chapter also discussed storage resource management. Viewing storage assets and the storage it consumes helps determine costs against the resources spanning certain business lines. Not only do enterprises need traditional storage resource management, it must be combined with a SAN management platform that encompasses the entire infrastructure.

Bibliography

Booch, G. (1994) *Object Analysis and Design with Applications* (Addison-Wesley Object Technology Series), Addison-Wesley

Booch, G., Rumbaugh, J., Jacobson, I. (in press) *Unified Modeling Language User Guide* (Addison-Wesley Object Technology Series), Addison-Wesley

Maksimchuk R, Nailburg E. (2005) *UML for Mere Mortals,* (Addison-Wesley for Mere Mortals Series), Addison Wesley

World wide web resources

LBL Technology Partners (http://www.drplan.com/business impact.htm) offers business impact consultancy and services that assist companies in improving their financial position.

The Object Management Group (http://www.omg.org) is the main source for the formal UML documentation.

Rational Software Corporation (http://www.rational.com/uml) offers a wide variety of information pertaining to the UML.

Ross Mayfield's Weblogs: Markets, Technology and Musings http://radio.weblogs.com/0114726/whitepapers/Timing your Business Case-Ross Mayfield.pdf

11
Integration

What exactly is information? Is it a scientifically useful idea? Can it be measured? Will it yield mathematical analysis? Therefore, to understand it we must first define it, but before we can define it – we must first describe it.

11.1 Understanding compliance requirements

The rise in data volumes has seen corresponding tightening of corporate governance and legal procedures surrounding data retention and availability. According to one large storage vendor, over 4000 major regulations apply to keeping information worldwide. The USA has the most procedures, with federal statutes, such as HIPAA, covering medical records, and Food and Drug Administration Section 21 rules that levy heavy fines for data retention noncompliance. The most commonly quoted USA regulation is the Sarbanes-Oxley Act of 2002, introduced following the Enron debacle where allegedly important documents were shredded. In the UK, there are extensive rules concerning the keeping of public sector information for extended periods. In the USA, the Securities and Exchange Commission (SEC) also has extensive rules around data retention with particularly heavy fines and even jail sentences for executives in instances of noncompliance.

The UK's Public Records Office and the USA National Archives and Records Administration (NARA) are two examples of institutions whose entire business is focused on ensuring records are

---

Data Lifecycles: Managing Data for Strategic Advantage   Roger Reid, Gareth Fraser-King and W. David Schwaderer   © 2007 VERITAS Software Corporation. All rights reserved.

maintained correctly and effectively. Most countries worldwide have similar governmental structures to ensure data is archived and released within a legal framework. These guidelines must be adhered to by other governmental agencies such as social security administrations or internal revenue services, as well as businesses manufacturing or trading within those countries. This can make things confusing when an organisation is based in one country, manufacturing in another and trading in another 40 countries around the world. And there seems to be no let up in the amount of new legislation coming into law. In Europe, new EU legislation will force telecommunications companies and Internet Service Providers (ISPs) to retain information on customer phone call logs or email and Internet connections, beyond the one- or two-month period that the information is normally held for billing purposes. The period could be up to a year and will supposedly help assist police and fraud investigations.

Compliance is forcing companies to rethink their migration strategies, not only fearing Hierarchical Storage Management (HSM) might be too simplistic, but also acknowledging that HSM can no longer meet the retention stringencies most laws demand. Looking at the effects of industry regulations as well as the availability and affordability of disk-based type protection, a number of observations are apparent with respect to the Data Lifecycle Management process, the types and timing of media used, the method of discovery and recovery applied and the convergence between various functions. As mentioned in earlier chapters, different regulations require different levels of retention and data protection. Therefore, a best practice is to understand compliance requirements clearly.

Outlining business requirements is one of the foundational concepts discussed earlier. Using models, the business requirement (or set of requirements) can help validate the solution used to fulfil the requirements. It is said that these requirements are the contract between IT and the business stakeholders that specifies how the solution should support the business's data lifecycle strategy. In most cases, the requirements will have two high level categories:

- function requirements;
- non-functional requirements.

Functional requirements are the general requirements. For example: 'The solution should provide the capability via policies to keep archived data on two or more separate managed offline media for a period of 1–7 years.' The non-functional requirements are collective requirements such as performance and extensibility.

That said it is essential for companies to evaluate their business quickly to first understand the real compliance requirements. This allows them to develop, architect, propose and implement a solution or a set of solutions to address this still-emerging trend. Information knows no boundaries and, where there is information, a compliance regulation will be close behind, more likely than not. The following data classification process will help prepare enterprises to meet this goal.

The most persistent conclusion evident from examining a small number of regulations is that all regulations are in constant flux. Hence, we are now witnessing an initial onslaught with respect to compliance impacts. Moreover, all indications are suggesting compliance will remain a primary discussion topic for CIOs for years to come.

According to the Enterprise Storage Group report (Gerr et al., 2003), there are four types of technology segments and media types that satisfy a majority of the compliance requirements:

- discovery;
- legibility;
- auditability;
- authenticity.

Arguably, securing your assets is a necessary step in this process. In any case, discovery is the first of these requirements. Enterprises must be able to discover data assets in a timely fashion. Or, better yet, they must know what they have before they can act upon it. Second, they have to know the data is there for a purpose. Without purpose, it doesn't need managed care. Third, enterprises must be able to provide an audit trail of the data once it is discovered, moved or altered. Last, enterprises must be able to prove its authenticity. Any solution must address all of these requirements to manage the new data lifecycle effectively in the coming years. Following discussion touches upon each of these requirements.

Recall the previous modelling discussion so that we can elaborate further on this important topic. Consider how models could be derived and used for compliance purposes. Given that the world is being controlled by versions of the Sarbanes-Oxley Act, let's look at that example specifically. The Sarbanes-Oxley Act requires company executives to attest to financial report accuracy. It is said that the general ledger system is the source for such financial reports. However, there might be multiple general ledgers, particularly in cases of corporate mergers. But, there is an opportunity of securing one source of truth as Louis J. Eyermann III (2004) discusses in an excellent paper: 'Protect Your CEO and CFO – How MDA Can Help Your Company Comply with Sarbanes-Oxley and Develop One-Source of the Truth!' In his paper, and according to the Hackett Benchmarking Group, most corporate finance groups spend about 20 % of their time gathering data, 20 % reconciling it and 30 % consolidating it. However, since Sarbanes-Oxley, companies now spend another 10–15 % reconciling the reconciliations. That only leaves around 5 % for distribution and 5 % for analysis. Can you believe it? Only 5 % left for true analysis?

Eyermann further notes that in a September–October 1994 Harvard Business Review article, Peter Drucker said: '. . . most companies do not understand *The Theory of the Business*. Expressed in more contemporary terms, corporate leadership does not understand their business model.' He further mentions that '. . . there are no systematic and agile IT processes in place to solve their current Sarbanes-Oxley problems let alone to synchronise the Sarbanes-Oxley solution as their business model changes. In short, they are reacting by putting band-aids on the problem.' These Band-aids could be a hardware or a temporary software fix. As you know, band-aids also fall off.

Now, this is where models become important. The Model Driven Architecture (MDA) developed by the Object Management Group (OMG) is a collaboration and design approach targeted at creating a set of agile and cost effective IT solutions to business problems such as a Sarbanes-Oxley solution. The MDA is a methodology and a standards collection for authoring, transforming and managing specifications that apply to any company's theory of business: its processes, systems and architectures. The methodology is both generic yet comprehensive enough to support principled design, development, testing provisioning, perpetuation, evolution and lifecycle management of systems. By treating all computing, including that of software and hardware, as a *system* of interrelated specifications, the MDA adapts well with various applications, architectural approaches (with its principled machinery for the authorship), transformation and management.

The MDA and its methodology provide an outline on how to separate and model business rules into platform independent models (PIMs) which are not coupled with the technical environments required of platform specific models (PSMs) that execute operations. The Computation Independent Model (CIM) expresses objective, function, process, policy and constraints of a system in noncomputational terms, e.g. domain, business rules and business vocabulary. This is easier for nontechnical managers to understand. However, this discussion is less concerned with CIM than PIMs and PSMs.

Once the architectural model is developed through collaboration with certain information brokers (subject matter experts) within the enterprise, the MDA gives a broader scope of tools over the general UML tools by using the Enterprise Distributed Object Computing (EDOC) compliant tool set. This is just an extension of the UML with a new UML diagram that is actually a culmination of multiple UML diagrams. Therefore, the EDOC is a great application for Service Oriented Architectures (SOA), Business Process Management (BPM) and Enterprise Architecture. Since BPM is the foundation for financial planning processes such as budgeting, forecasting and reporting, it is ideal for effectively implementing a Sarbanes-Oxley plan, but not just that alone.

You may not have heard of the EDOC specification yet, but you should have. Although EDOC may not be well known, it is only a UML extension. Therefore, it is gaining notoriety, but might not be well understood despite being great at addressing financial reporting processes and developing compliance ready solutions, creating links to legacy systems and importing and correcting data from source files and various other feeder systems in an environment. And that is what we are discussing here – integrating a lifecycle management solution(s) with processes and methodologies that can help you and your enterprise.

In order to develop a compliancy procedure the first step is to model the Sarbanes-Oxley process. In this process, all of the financial system relationships should be modelled. This reveals the systems and their processes. The second step is to link these with technical platforms and map the PIMs to the PSM and then map them down to configuration, documentation and deployment. Note that the PIM should be connected with as many Lines of Businesses (LOBs) as possible. This is where one can start mapping out the specific processes and requirements discussed earlier. There are many more steps involved, but they are omitted for brevity purposes. This allows a quick glimpse of using models for integrating solutions, remembering that lifecycle solutions are composed of technology, processes and methodologies which should include UML and EDOC.

11.1.1 Automating data lifecycle management

Data Lifecycle Management (DLM) can be described both from an enterprise customer and an overall solution perspective. A DLM target solution focuses on integrating existing and future data management products, which, coupled with workflow management, provides the tools, infrastructure and services for enterprise corporations and small and medium businesses to operate their respective DLM processes. A DLM solution provides an integrated and automated approach to data availability, data protection, data retention management and data integrity while maintaining application transparency and end user uniformity, including common federated data classification and policy management, as well as an integrated management paradigm. Reiterating, lifecycle management is not a single technology nor is it only a service based process; it is an intertwined relationship between technology products, processes and methodologies. This combination represents an entire data and information integrity management platform.

11.1.1.1 Hierarchal storage management: the first wave of data lifecycle management

In many ways, DLM represents the evolution of HSM techniques. Vendors first developed HSM products in the 1980s in the mainframe environment for distributed computing implementations. HSM offers several benefits. It reduces the total amount of expensive RAID disks an enterprise needs and increasing storage-use efficiency can improve performance. Moreover, enterprises can perform various routine storage housekeeping tasks more easily with HSM products.

For decades, HSM has existed in the mainframe arena. HSM was initially intelligent software that moved data automatically from expensive media to lower cost storage by rules the business imposed. Organisationally isolated administrators developed policies to search and sweep storage based upon such considerations as file creation time, modification time, access time, location, size and name. The files then migrated to alternate storage based on those criteria. Data could be purged from the original location or at a specific time. When purged, the HSM application would insert a file *placeholder* that allowed users or applications to access the file seamlessly regardless of its physical location. We say seamlessly, but if the file had to come from a tape medium, the process could take noticeably longer than coming from an alternate disk location. Because of the policy based system, it was easy to manage the lifecycle of the data.

In HSM implementations, data automatically moves from expensive hard disks to less expensive optical media or to tape according to specific policies. Users don't have to know their data has migrated to a less costly storage media because HSM products track data movement and create data retrieval paths. When an HSM product moves data, it creates a pointer to a file's new location. When a user or application retrieves a file that has moved down the storage hierarchy, the HSM product automatically returns the data to the top level of the storage infrastructure. Companies that use an HSM approach typically use two *triggers* to move data. The most common trigger is time. Data that workers haven't used within a specific time period moves to a less expensive storage device. The second trigger is capacity. As disks fill, data can move down the hierarchy. What gives HSM the capability for this movement is the Data Management Application Interface (DMAPI) specification which, when implemented by a file system, enables HSM engines to manipulate file system metadata, and migrate and retrieve data as required by various policies.

Since inception of HSM into the marketplace, most implementations supported UNIX servers. However, Microsoft introduced the concept of *reparse points* in Microsoft Windows NT file system, version 5, that allows automatic movement of data based on objects called reparse points. This provides the ability to create special file system functions and associate them with files or directories, and enables the file system functionality in NTFS to be enhanced and extended *dynamically*.

In most cases, using reparse points begins with applications. Applications that use reparse points are generally Backup and

Archive applications. So, when an application wants to use the feature to store data specific to the application, which can be any sort of data at all, it stores it into a reparse point. The reparse point is tagged with an identifier specific to the application and stored with the file or directory. A special application-specific driver filter is also associated with the reparse point tag type and made known to the file system. More than one application can store a reparse point with the same file or directory, each using a different tag. Microsoft also reserved several different tags for its own use.

Suppose that a user decides to access a file that has been tagged with a reparse point. When the file system goes to open the file, it notices the reparse point associated with the file. It then 'reparses' the original request for the file, by finding the appropriate filter(s) associated with the application that stored the reparse point, and passing the reparse point data to that filter. The filter can then use the data in the reparse point to do whatever is appropriate based on the reparse point functionality the application intended. It is a very flexible system; how exactly the reparse point works is left up to the application. The really nice thing about reparse points is that they operate transparently to the user. You simply access the reparse point and the instructions are carried out automatically. This creates seamless extensions to file system functionality.

In addition to allowing reparse points to implement many types of custom capabilities, Microsoft itself uses them to implement several features within Windows 2000, including the following:

- *Symbolic linking* allows creating a pointer from one area of the directory structure to the actual location of the file elsewhere in the structure. NTFS previously did not actually have symbolic file linking. However, newer systems, and especially Vista will, but the functionality can be simulated by using reparse points. Therefore, the symbolic link is just a reparse point that redirects access from one file to another file.

- A *junction point* is close to a symbolic link, but rather than redirecting access from one file to another, it redirects access from one directory to another.

- A *volume mount point* is similar to a symbolic link or junction point. Its primary function is to create dynamic access to entire disk volumes. For example, you can create volume mount points for removable hard disks or other storage media, or even use this feature to allow several different partitions (C:, D:, E: and so on) to appear to users as if they were all one logical volume. These could be used as folders on Windows 2000 and 2003. Moreover, you can use this capability to break the traditional limit of 26 drive letters – using volume mount points, you can access volumes without the need for a drive letter for the volume.

- *RSS Server.* Beginning in Windows 2000, the RSS Server feature uses a set of rules to determine when to move infrequently used files on an NTFS volume to archive storage (such as CD-RW or tape). When moving files to 'offline' or 'near-offline' storage in this manner, RSS leaves behind reparse points that contain the instructions necessary to access the archived files, if they are needed in the future.

These are just a few examples of how reparse points can assist automating lifecycle management. As is evident, the functionality is very flexible. Reparse points are a nice NTFS addition – they allow file system enhancements without requiring change to the file system. Upcoming Windows Vista releases will exhibit even more functionality.

11.1.2 Content searching

So, Mr. Gates was right. Content is becoming all important and is at the forefront of CIOs and IT Managers' minds. In his famous essay of January 1996, Bill Gates wrote: 'Content is where I expect much of the real money will be made on the Internet, just as it was in broadcasting.' He went on to say: ' . . . I expect societies will see intense competition – and ample failure – as well as success in all categories of popular content' not just software and news, but also games, entertainment, sports programming, directories, classified advertising and online communities devoted to major interests.'

But what value does content offer if you cannot find what you need? Breaking news: Information retrieval will be the next hot topic! Today's organisations generate and store vast amounts of information. Whether it is corporate documents, databases, certain reports, email or other information, organisations capture it and save it in distributed repositories. But simply possessing the data and information does not automatically mean competitive advantage. To ensure success, corporate users need to find the right information they require easily and rapidly. This means that organisations wishing to exploit their knowledge assets fully require fast and efficient solutions for searching, browsing and viewing information. Corporate databases and various file systems – UNIX or NTFS – must have the ability to be searched quickly and efficiently.

There are two different categories of retrieval:

- data retrieval;
- information retrieval.

Content retrieval normally comes under the second category. Moreover, information retrieval is different than data retrieval, which is a more deterministic process where the goal is an exact match. In information retrieval, exact matches are great, but users would generally rather have *fuzzy* matches which find items most closely resembling the match and then choose the best from the results. This makes the searches more probabilistic. Having a search return more probable results is the reason for using Bayesian logic which is based on Bayes's Theorem.

Thomas Bayes, an 8th century English cleric, is known for his mathematical probability work that mostly centred around calculating the probabilistic relationship between multiple variables, determining the extent to which one variable impacts another. That said, there are additional techniques used to inference unstructured data which include:

- Clustering algorithms. Clustering algorithms gather information on similar documents and their connection to others. An example of this would be compliance solutions applications searching through email and other documents.
- Syntactic and semantic analysis. This technique can be used for searches against common terms.
- Ranking based on popularity of documents and user feedback. A prime Internet example is Google's html document search engine.

Search technology is here and has significantly improved over the few years of its existence. The new trend combines multiple search techniques in an entire bundled solution. In addition, emerging search technologies provide the capability to develop user profiles to assist in profiling individual searches.

Employing these powerful mathematical models and algorithms, companies are beginning to use engines (built on their own or by OEM partners) to deliver powerful information search and retrieval applications for document or content management applications. One of the companies providing these engines is Hummingbird.

In addition, backup and recovery software will soon be using advanced search engines to search for backups that are beginning to hold more and more data. This will give end users more powerful data and information retrieve capabilities.

By using content indexing, administrators can facilitate searching data from particular data storage pools or stores. Most content searching engines provide the ability to

- index data;
- carry out advanced searching.

Software products that employ indexing engines can index archived file contents and data stores messages. After files and messages are indexed, administrators can search for text words or for other attributes such as a message recipient or a file type (for example, *.doc). One could employ other searching methods such as Wildcard or Boolean searches. With this type of searching, administrators can use pattern matching to search in the 'Containing the word or phrase' field or in the 'All or part of the name' field. Typically, content indexing engines use Adaptive Probabilistic Concept Modelling (APCM) to rank the results that match a Boolean query.

APCM has been defined as

> *Analyses correlations between features found in documents relevant to an agent profile, finding new concepts and documents. This is an extension of the Bayesian theory. Using this technique, concepts important to sets of documents can be determined, allowing new documents to be accurately classified.*

> (Top Quadrant, 2003)

Using techniques like *clustering algorithms* and *fuzzy matching strategies* allows powerful searches. Consider a few search examples. In the first, administrators use natural language searches, which look for text that is similar, but not identical, to text you supply. For example, you could type *Exchange system agent* and receive a list of files that contain the words *Exchange Agent*. Natural language queries make it possible to find results without being familiar with search algorithms or syntax. Documents are ranked in the results by relevance. In a second example, administrators use exact phrase searches, which look for the identical characters typed within quotations marks. For example, you could type 'home office' to find files or messages that contain those two words separated by one space. If you want to search for multiple strings, put quotation marks around each string and separate them with plus symbols (+) or with spaces, as follows: 'home office' 'ergonomic chairs' 'full spectrum lighting'.

Note that latter example does not return results for similar phrases, such as 'ergo chair' or 'full-spectrum lights'. To search for similar patterns, use a search method other than exact phrase searching. Table 11.1 gives examples of Boolean searches that a majority of the search engines offer.

Table 11.1 Example Boolean searches

Operator	Description	Example
AND	Binary operators ensure that both terms are matched	**home+AND+office** Returns documents that contain both home and office.
NOT	Unary operator ensures that the term following NOT is excluded	**office+NOT+home** Returns only documents that contain office but not home.
OR	Binary operator makes sure one or both must appear in a search	**home+OR+office** Returns documents that contain home, office, or both home and office.
EOR & XOR	Another binary operator. This is an exclusive OR, meaning that only one of the terms is permitted to appear in the search	**home+XOR+office** Returns only documents that contain home or documents that contain office. Documents that contain both home and office are not returned.
()	This is a bracketed expression and is often used from left to right in a search	**(home EOR office) AND (ergonomic EOR chair)** Returns only documents containing one of the following combinations: home and ergonomic, home and chair, office and ergonomic, office and chair

As data retention continues for longer intervals and as the volume of data continues, there must be a way to search through the tremendous amount of data. These powerful and unique search engines will be paramount to successful lifecycle solutions. In selecting the most appropriate lifecycle tools with searching capabilities, there are certain factors to consider including:

- organisational readiness to engage in the continuous processes and methodologies necessary to achieve success for using search engines;
- format of search results;
- type of techniques used in the engines – e.g. Bayesian Logic.

11.2 Understanding hardware and its constructions

Before implementing data or information lifecycle technology, it is necessary to understand the hardware requirements and how the hardware operates since there are limitations in most data lifecycle and information lifecycle technical products. The use to which a data or information lifecycle technology is applied plays an integral role. More than likely, enterprises will implement the data or information lifecycle technology into an existing environment and that means it has to integrate with legacy hardware and software. There have been several instances where a purchased product did not match the initial project requirements. It might have not been the fault of the salesperson, but perhaps the fault of the team in determining initial requirements. Remember, DLM and ILM are relatively new fields and require strict attention to business requirements. More often, a solution is implemented for requirements purposes, or to report on the usage of storage across the enterprise to save money.

Another important hardware consideration is its adaptability. Will it be able to adapt quickly with future software? Some questions you should ask are:

- Is the hardware used by major software vendors?
- Is there a compatibility matrix?
- Does the hardware use open standards and APIs?

Hardware is expensive and more expensive in the long run if it does not scale or cannot adapt or be repurposed. Begin thinking outside the box. Just because an initial role was determined earlier, there shouldn't be a reason to prevent it being used in different roles. For example, just as tape was initially used for simple backup purposes, new tape roles have emerged such as long preservation archival of corporate data (e.g. HSM). In large environments, it is essential that vendors, integrators and the entire IT management team understand the purpose of the hardware. As more demanding software applications are deployed, a hardware platform that can adapt, scale and perform can only contribute to the overall success of the mission and its objectives.

11.2.1 Current storage technologies

This preceding discussion underscores an important and fundamental point. With the multiple techniques, strategies, methodologies, software and hardware for enterprise storage management, no single solution is pervasive. There will be different hardware and software in almost all corporate environments. It is also evident that different storage technologies and strategies will be used. In most cases, newer strategies are used in the backup and archival space. Backup has changed over the years as explained in Chapter 8 on continuous data protection. Disk has become more prominent in data centre environments, and is being intertwined with legacy tape backup strategies.

11.2.2 Disk-based storage strategies

Conventional data management and protection techniques have not kept pace with the increasingly complex nature of today's data processing and topology. Many IT organisations still use the 'weekly full, daily incremental' backup techniques deployed in the 1950s. In the past half century, topologies have evolved from centralised homogeneous platforms to heterogeneous networks of distributed and mobile systems and multiple storage tiers. Moreover, annual data growth has increased from 20–35 % to 80–100 % and retention periods have increased from weeks to decades. Finally, usage patterns have morphed from transaction to transaction and reference. Studies indicate that approximately 60 to 80 %

of this growth is fuelled by reference data that describes information with access requirements measured in seconds to minutes while transaction data access is in milliseconds.

Traditional data protection and management approaches are doomed to inevitable failure in today's complex environments. Ernst and Young's *Fabric of Risk* study determined that approximately 36 % of the executives from the top 1000 publicly-traded companies believe their companies would cease operations due to inadequate protection, while 59 % place their risk as moderate to high. The increased dependence upon networked and mobile data, combined with theft and viral infection vulnerabilities, exacerbates this risk. New technologies such as SANs aid physical storage management, but affordable data protection and management solutions have proven elusive indeed.

As many studies and onsite audits point out, remote site data backup is mostly executed by non-IT personnel. Sometimes, the wrong tape is inserted and no one else is present to check. The response to system errors is often incorrect and is not reported to a central IT authority. Even worse, when a tape loading process fails, no backup is executed. As the 'remote' employee is often not qualified to verify whether the backup has been successful or not, no one really knows if the tapes contain the correct data.

The massive growth in data generation and retention periods combined with legislative requirements requires a fundamental change in backup procedures. That said, new disk-based strategies and technologies have to be drawn. There are technologies that provide content routing, a technology that provides a long-term framework to address both the rapidly growing volume of data and the wider information lifecycle issues. There are software solutions which have a specialised disk feature for the protection of file data on clients, anywhere on the network to any type of disk-based storage pool.

With this unique distinctive fingerprint technology, called global unique file identification, the disk distinguishes unique files from redundant copies across the enterprise. Enormous storage capacity and network traffic savings are achieved by not transmitting and storing redundant data.

A backup across three remote office file servers each having the same two megabyte Word files would result in transferring six megabytes with conventional approaches to backup. A backup application using with this disk feature such as this, however, stores a single copy only and consequently needs no more than two megabytes of storage capacity – a 66 % savings as compared to the space that other solutions require. Comparable results are seen for throughput as the backup would complete about 66 % faster across the three remote sites.

Through file segmentation, the distinctive fingerprint technology can even be applied on small parts of files for global unique segment identification. This is typically done for large files like .pst files. Furthermore, it is clear where this could be used with a combined lifecycle scenario. When such a file is modified after the previous backup, only modified segment(s) will be backed up.

Global unique file and segment identification is performed by lightweight client systems disk agents. Centrally, backup data is stored in a Disk Storage Pool. The Metabase in the Disk Storage Pool stores the file metadata in a scalable distributed database. The file content segments are stored in one or multiple Content Routers in the Storage Pool. As metadata and content data are stored separately, all source file versions can be restored, while only the globally unique file segments are stored.

In a typical system, about 13 % of the files are modified each day and must be backed up. Using a disk global unique file and file segment identification, studies show the number of actual bytes changed is 1–2 %. Take for example 1 TB of data. When doing four incremental backups and one full one each week over a period of 16 weeks, the 80 backups require more than 16 TB of storage. The disk stores an optimised initial backup and then 16 weeks of incremental backups. After 16 weeks, the disk typically needs 1.5 TB to store all versions of all files within this retention time. With 80 TB of source data protected in 80 backups of 1 TB, the disk has reduced the backup data volume by over 50 times the amount of data.

This book isn't really focused on describing data backup, so further disk strategy discussion will be centred on using virtualisation and file system technologies to move and group data for lifecycle management. That said there are alternative ways to group data and therefore make it more organised. For example, certain virtualisation software companies support logical volumes as a single logical object. All I/O to and from an underlying logical volume is directed by way of this object. Using such techniques allows the mapping of more than one device to a single file system. Administrators can then configure policies that automatically relocate files from one device to another, or relocate files by running file relocation commands. Having multiple devices lets administrators determine where files are located, which can improve performance for applications that access specific types of files. The feature also allows file systems to reside on different classes of devices, so that a file system can be supported from both inexpensive disks and from expensive arrays. In this manner, administrators can control which data goes on which volume types as per the business requirements. For example, they could configure policies that automatically relocate files from one device to another, or they can relocate files by running file relocation utilities as commands. One can think of this as of having two distinct processes:

- *Relocation policies.* Policies that administrators configure to determine which files to move to different component volume, utilising this as a technology step from an earlier defined model.

- *File relocation.* The relocation utilities recursively examine a multiple-component file system searching for files matching configured relocation policies. After searching a file system, the utility passes information about files that match policies to an additional move utility. A move utility reads a list of file names and locations that the relocation utility has created and relocates files based on that relocation list.

11.3 Understanding user expectations

Data lifecycle management gives administrators a framework to understand the value of different records and helps them build storage infrastructures that reflect those determinations. A more enlightened approach to data management comes from understanding the lifecycle of data and applying this knowledge to ensure data resides on resources that deliver appropriate quality of storage services. In other words, by understanding dataset utilisation patterns and retention requirements, IT departments can route data to resources with appropriate levels of performance, protection, availability, retention, immutability and cost. However, technical administrators need to understand user expectations before they can use a DLM approach to assemble the proper combination of storage devices, media types and network infrastructure to create an appropriate balance of performance, data accessibility, easy retrieval and data reliability based on the relative value of the data. Users should know that any lifecycle process and technology evolves. There will also be an ongoing process for gathering requirements and releasing functionality within the technology pieces of a data lifecycle solution. However, the models that were built will be key ingredients in reducing the time associated with gathering the requirements. Since the DLM approach examines data capture, transfer, processing, analysis, storage, backup, retrieval, archiving and deletion it covers immense ground. In using this 'user expectation approach,' administrators determine whether they need to store data online, near-online or offline, and when data should be deleted. All the previous steps, from building the initial business models to the classification process, allow the lifecycle process to unfold (Figure 11.1).

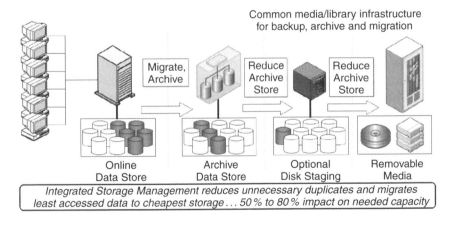

Figure 11.1 Lifecycle Management Process

11.3.1 Organising data

We have discussed classifying data, but we need to relate that to information retrieval and how end user expectations can be fulfilled. Typical corporate users reference documents and other information daily, perhaps in the form of email or documents. Presently, many vendors seek to be the leading player in marketing automatic data classification engines. These vendors are providers of enterprise software that enables organisations to automatically understand and harness critical business information. These providers supply enterprises of all sizes with the ability to automatically unify, manage and utilise information stored on their systems and within the workforce to deliver competitive advantages such as reduced business cycles, accelerated productivity and an increase in revenue. These automatic engines organise and then classify the data. In addition, they fully automate the processing of any type of unstructured information, i.e. text, audio and video. Technology such as this can automatically categorise data with no requirement for manual input whatsoever. Categorisation feature flexibility allows administrators to derive categories precisely using unstructured text. This ensures that all data is classified in a correct context with utmost accuracy.

Today, most leading companies use a combination of the various techniques mentioned earlier. Their strength lies in advanced pattern-matching techniques (nonlinear adaptive digital signal processing), rooted in the theories of Bayesian Inference, which enable identifying natural text patterns based on the usage and frequency of words or terms corresponding to specific concepts. Based on the preponderance of one pattern over another in a piece of unstructured information, these advanced software engines enable enterprise computers to infer there is a particular probability that a document in question discusses a specific subject. In this way, software can extract a document's digital essence, encode the unique 'signature' of the key concepts, and then enable a host of operations to be performed on that text, automatically. Most businesses immediately see the benefits of full metadata and XML handling, including weighting, and many functions and field types (such as financial and date), as well as the ability to do combinatorial operations. As companies gravitate to using these types of techniques, a clear understanding of the value they bring to the business soon becomes evident. But it does take effort on behalf of the company to implement and integrate the capability into their environment.

Well-trained tacticians can see it – organised and classified data becomes information: information that is located quickly and efficiently comprises a recognised asset to the business. In the end, data retrieval transforms data into *Living Data*. From

Administrators, Storage Practitioners, Architects, Engineers and business minded individuals, sophisticated software does require effort investment. As this discussion suggests, this technology is being embedded in traditional backup and archiving software and in applications addressing many markets, including:

- business intelligence;
- content publishing;
- document management;
- e-commerce;
- electronic customer relationship management;
- email routing;
- records management;
- marketing and sales automation;
- portals;
- security.

Here are some points to ponder when considering software that uses some of the technology this section mentions:

- Does the technology scale?
- Does the technology support XML and metadata searches?
- Can users edit or change the way the technology searches for information?
- How does the indexing functionality work?
- Is security a part of the technology?

11.4   Knowing the capabilities of your data management tools

Just as a carpenter knows the capabilities of a saw and hammer, administrators should be familiar with their data management tools. Their tool belt can be large, but each tool has different functions or a combination of functions. Now consider another essential data management tool – new virtualisation technologies. Virtualisation is presently playing an integral lifecycle management role because data is held on file systems or groups of disks and requires organisation and proper safeguarding by various RAID techniques. However, where the software should reside has been a hot discussion topic in the storage industry.

11.4.1 Virtualisation of storage, servers and applications

Network-based volume management has emerged in the past few years and, to many, it represents the true potential of virtualised storage in heterogeneous environments. Broadly speaking the rationale is that moving volume management into the network enables creating a storage pool that is independent of both the host and the back-end storage. Residing on either dedicated appliances or on next-generation *intelligent* storage switches, proponents argue this approach can deliver more than just storage pooling, with centralised storage management, advanced storage services and even superior throughput performance among the benefits. However, there is no present consensus on how this is best implemented – either on the host or the network. As this relates to lifecycle management, there are points to be made for both sides, but having virtualisation technology is essential for effective lifecycle solutions. However, there are some necessary questions for when administrators place this process at the switch layer as shown in Figure 11.2:

- Does this place the infrastructure at risk?
- Does host based virtualisation cause major disruptions?
- What disruptions could occur with it installed at the switch layer?
- Are staff sufficiently trained to understand the nuances of this new technology?
- What other companies are utilising this new technique?
- Will the various virtualisation techniques integrate into a unified management layer?
- Will a vendor willingly show a roadmap?

It is sometimes thought that storage doesn't necessarily have a monopoly on virtualisation. In fact, the term *virtualisation* is almost ubiquitous, most often used in conjunction with on-demand/utility computing schemes that attempt to automate the

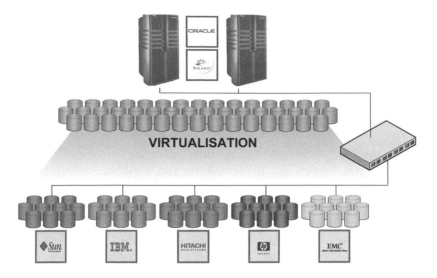

Figure 11.2 Bringing storage virtualisation to the network

management of systems, software and networking resources – as well as storage. Currently, Data Centres must have separate administrator teams for these individual elements, at great expense. The missing piece is the *one pane of glass management platform* as the glue for it all. And, all those highly touted virtualisation benefits – such as increased utilisation rates, resource consolidation, simplified management, automated provisioning and workflow administration – need to run across both the server and storage infrastructure before full gains can be realised.

Vendors and strategists still argue about how the various levels of virtualisation will exactly come together. And, while there are some encouraging signs of fledgling progress toward a standard method of storage virtualisation (through SMI-S), there is not much in the way of any standards. Just as storage intelligence can be distributed at various infrastructure points, virtualisation technology can integrate at different levels, from high in the application stack right down into a chip or networking hardware. These various approaches are suitable for different tasks, and sometimes they actually work together. However, working with various companies around the globe, it's evident that a general emphasis shift is occurring from static to dynamic computing resources, including virtual applications and databases, virtual private networks, virtual operating systems and virtual hardware.

11.4.2 Product technology and business management functionality

Storage management tends to be selected on a vendor-by-vendor basis, the result being that most infrastructures run a large number of vendor-specific management consoles. Not having a unified *one pane of glass platform* does not help businesses understand or know the state of their storage or IT infrastructure. In addition, storage capacity can be marooned causing difficulties in allocating storage freely between multiple host applications according to changing business requirements. Tasks, such as storage hardware migration, tend to be labour-intensive exercises that also impact application availability since applications must be stopped to perform migration. All of these tasks should be easily visible and reported on for business level reporting. A logical process could be viewed as in Figure 11.3.

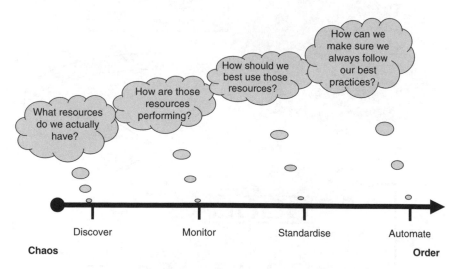

Figure 11.3 Management platform functionality process

11.5 Solution integration – business data and workflow applications

ITIL (the IT Infrastructure Library) is the most widely accepted approach to IT service management in the world. ITIL provides a cohesive set of best practices, drawn from the international public and private sectors. It is supported by a comprehensive qualifications scheme, accredited training organisations and implementation and assessment tools. The best practice processes promoted in ITIL support are supported by the British Standards Institution's standard for IT service Management (BS15000).

Therefore, what is the best model to achieve alignment of business and IT objectives? Where do you start and what are the steps involved?

- Operational efficiency is bringing chaos to order (see Figure 11.4).
- True service delivery is bringing critical isolated IT infrastructure into the visible and usable form: that is bringing isolation to interdependence.

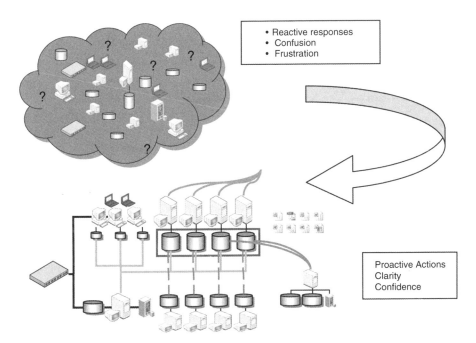

Figure 11.4 Chaos to order

Currently, most enterprises are isolated from their IT and the result is that they

- make critical decisions without assessing the impact of technology;

- notice their IT organisation only when something's broken;
- assume their ever-growing IT budget is spent inefficiently.

IT is similarly isolated from the enterprise, with the result that they

- make critical decisions without assessing the impact on business;
- plan resource requirements based on current usage, not future business opportunities;
- react to immediate problems rather than prioritising from a business perspective.

11.5.1 Standard management and reporting platform

A common platform should be available to the enterprise that provides an integrated solution of SAN Management and other entities listed below.

11.5.1.1 Storage and service management

- Visibility: Discover SAN, NAS, DAS; report, trend, forecast; monitor and alert on problems (see Figure 11.5).

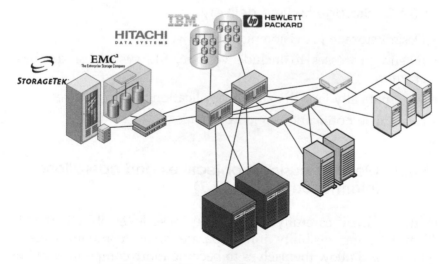

Figure 11.5 Visibility into storage

- Delivery: Storage provisioning workflow automation (see Figure 11.6).
- Accountability: Storage service levels, usage and chargeback.

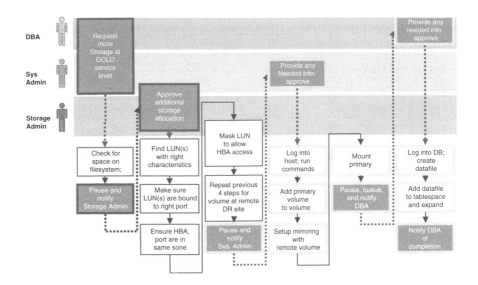

Figure 11.6 Automating lifecycle mangement operational

11.5.1.2 Storage service delivery

- Design storage provisioning workflow.
- Extend processes to include volume, file system or database creation.
- Automatically launch external applications during process.
- Customise notifications and approvals.

11.5.2 Meeting business objectives and operational information (Figure 11.7)

In the end, by carefully aligning business level objectives with IT and giving visibility into separate areas, companies create synergy and allow themselves to become more competitive in this new era.

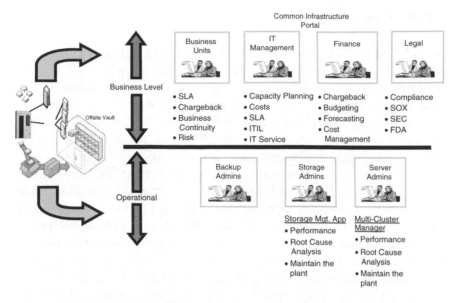

Figure 11.7 Business and operational synergy

11.6 A ten-point plan to successful DLM, ILM and TLM strategy

- Clearly define the projects' mission and objectives.
- Begin by understanding your business and communicating with models.
- Use models for data classification and business alignment.
- Use virtualisation techniques to group, move and map data efficiently.
- Consider archiving solutions that can be integrated with existing backup solutions.
- Consider a cross-platform data management platform for SAN, NAS, DAS, SRM and business levelling reporting and others.
- Consider backup solutions that offer advanced features and can easily integrate with compliance solutions.
- Maximise current employee resource training in lifecycle management solutions.
- Model Business Requirements.
- Start with a subset of the infrastructure and build upon success.

11.7 Conclusion

So is DLM just a departure back to the dizzy heights of HSM? No, of course not; although getting your data (and ducks) in line with its most appropriate level of storage makes absolute sense and in that respect there is some commonality. HSM plays an important role in tiered storage strategies, but is not what managing the lifecycles of data is about. For the first time we are now considering the worth, defence, preservation, accessibility and overhead cost of maintenance of preserving, storing, retrieving data and information across a lifecycle that spans from here to eternity. DLM redefines how we approach storage, data and server management. We used to focus on the storage infrastructure and how we maintain that infrastructure for data availability for disaster recovery or business continuity purposes only. DLM takes this concept further and fully exploits the alignment between IT and lines of business, together with the value that can be gained from that exploitation. No longer can we even consider deleting data based on age; the assumption that the oldest data had the least value. Data needs to be managed and moved and deleted at specific intervals, most definitely not based on time but based on its business value, at that specific time, together with the external requirements set by governments, industry bodies and policy management strategies. In order to avoid bestowing more costly storage to data that is no longer of any value and put it in the most appropriate storage takes a great deal more intelligence. DLM is more intelligent than previous iterations of storage, data, or information management. It involves data movement, archiving tools, hierarchical storage management and the ability to handle the cycle of hardware and software that is required to manage a single piece of data from cradle to grave.

References

Ernst and Young: Fabric Risk Study: http//www.ey.com/ global/Content.nsf/International/Canada_-_Publications

Eyermann (2004): http//www..omg.org/news/whitepapers/ OMG_Article_December-2004-MDA-and-Sarbanes-Oxely.pdf

Gerr, P.A., Babineau, B. and Gordon, P.C. (2003) ESG Impact Report: Compliance: The effect on information mangement and the storage industry. (2003). More information on the report can be found at: http://www.enterprisestrategygroup. com/Search.asp

Kosierok, Charles (1997) *NTFS Reparse Points*, Retrieved 24 January 2005 from http: // www.pcguide.com/ref/hdd/file/ntfs/files Reparse-c.html

Gates, William (1996) *Content Is King*, Retrieved 24 January 2005 from http: // www.microsoft.com/billgates/columns/ 1996essay/essay960103.asp

Top Quadrant Research (2003) *Dictionary of Search Termi-nology*, Retrieved 1 January 2005 from http://www. topquadrant.com/documents/TQTR-Search02.pdf#search= 'Adaptive%20Probabilistic%20Concept%20Modeling'

Index